Dr. Adrian Kaminski

# LABORATORY EXERCISES
# IN ASTRONOMY

Phoenix College
Physical Sciences Department

**AST 102**

---

# Astronomy 102 Laboratory Manual

---

**3rd Edition**

Dr. Adrian Kaminski

Order this book online at www.trafford.com
or email orders@trafford.com

Most Trafford titles are also available at major online book retailers.



Printed in the United States of America.

ISBN: 978-1-4907-3451-4 (sc)
ISBN: 978-1-4907-3452-1 (e)



*Trafford rev. 04/24/2014*

www.trafford.com

**North America & international**
toll-free: 1 888 232 4444 (USA & Canada)
fax: 812 355 4082

## — Contents —

# Northern Polar Constellations

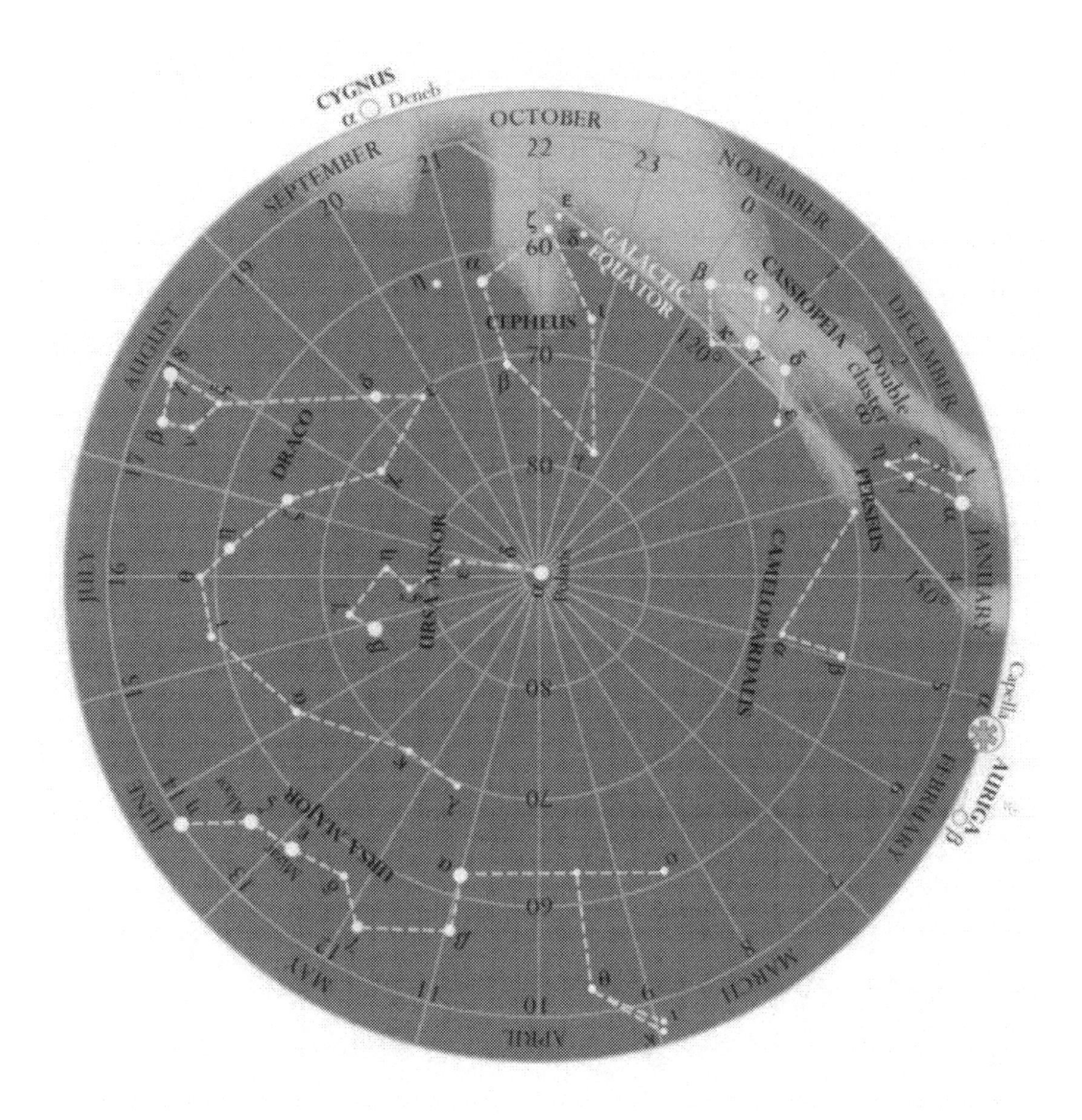

# Southern Polar Constellations

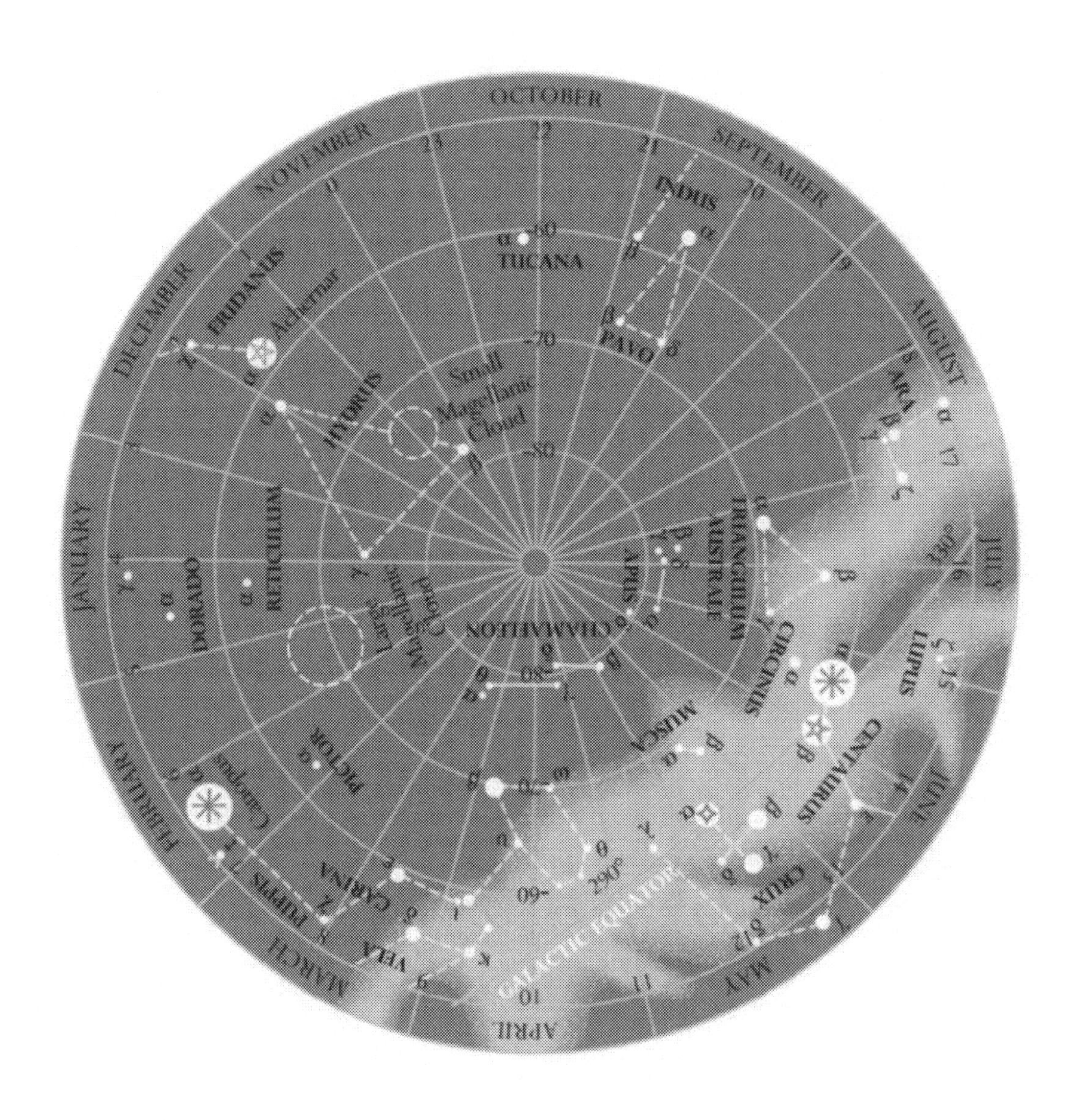

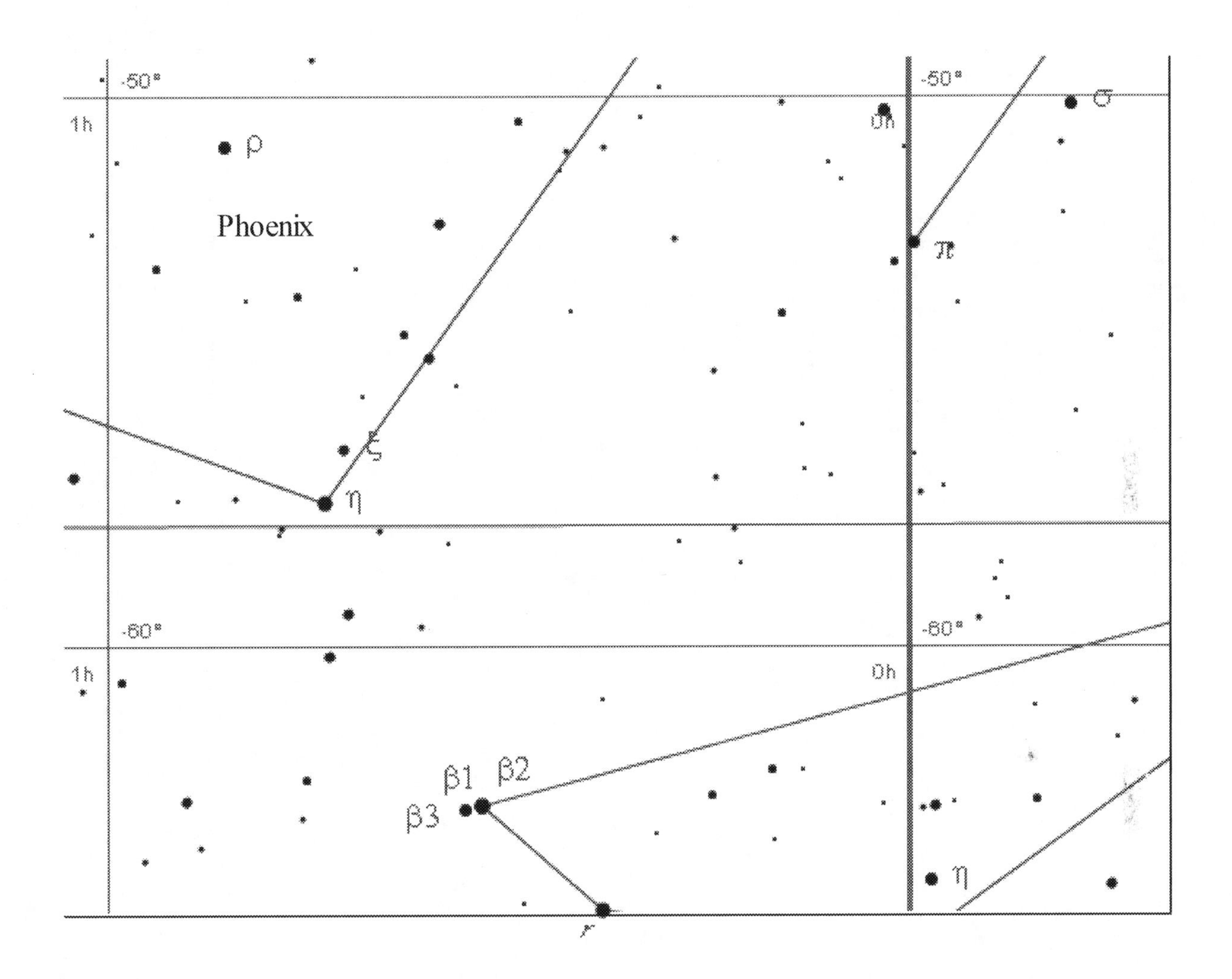
-50°
1h
ρ
Phoenix
0h
-50°
σ
π
ξ
η
-60°
-60°
1h
0h
β1
β2
β3
η

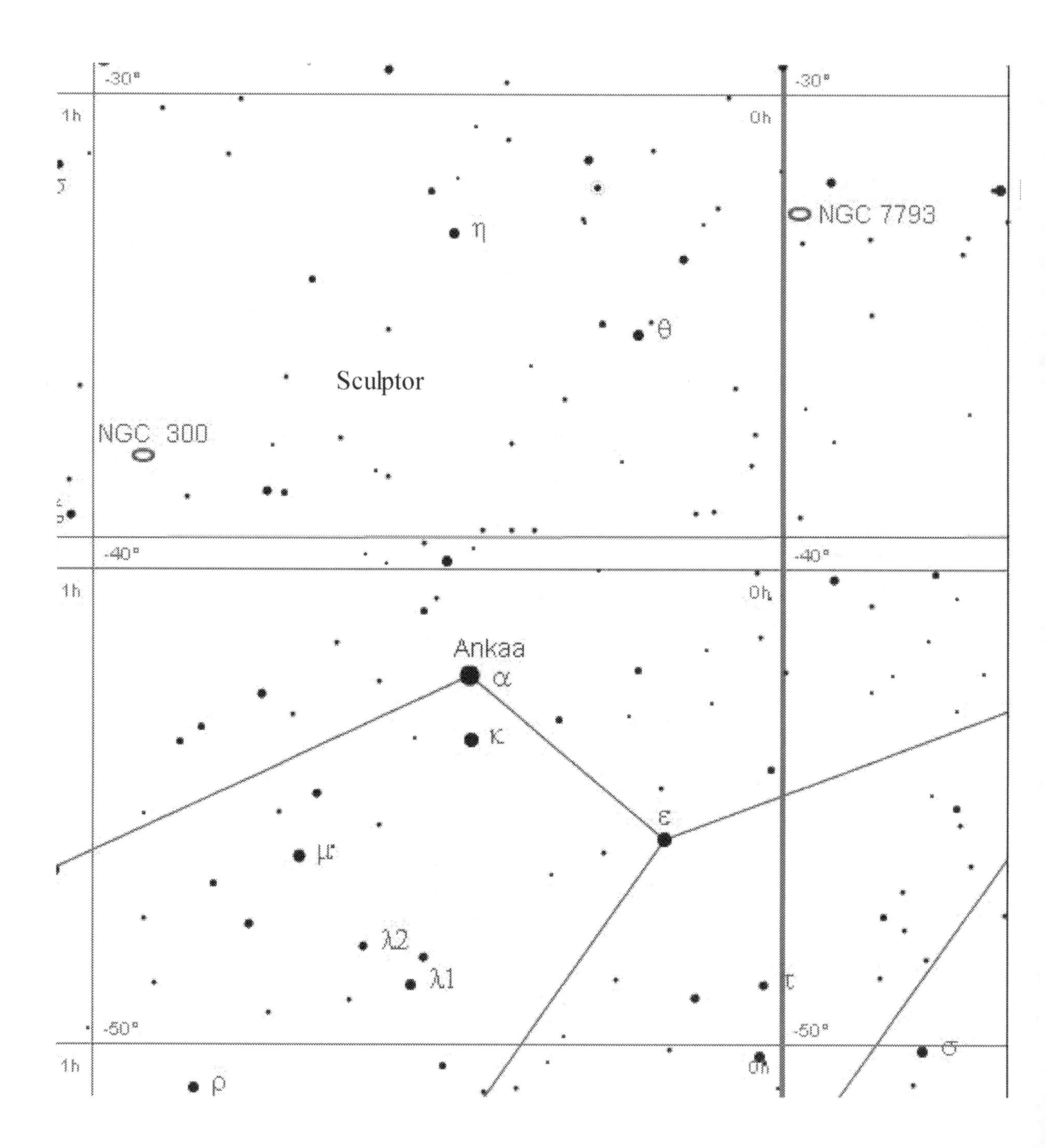
-30°
1h
0h
-30°
NGC 7793
η
θ
Sculptor
NGC 300
-40°
1h
0h
-40°
Ankaa
α
κ
ε
μ
λ2
λ1
τ
-50°
1h
0h
-50°
σ
ρ

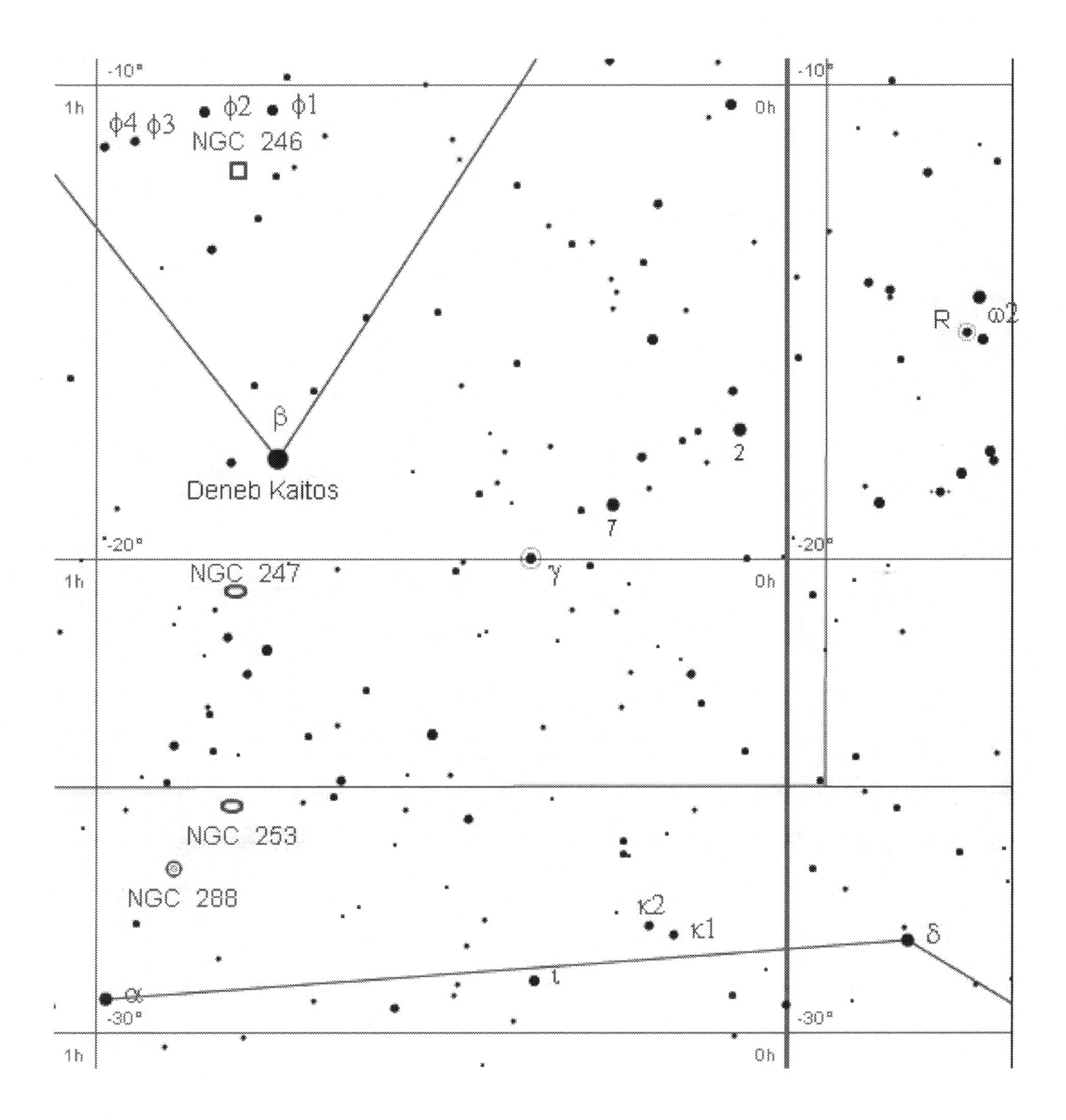
-10°
1h
φ4 φ3
φ2
φ1
NGC 246
0h
-10°
R
ω2
β
Deneb Kaitos
2
7
-20°
-20°
NGC 247
γ
1h
0h
NGC 253
NGC 288
κ2
κ1
δ
ι
α
-30°
-30°
1h
0h

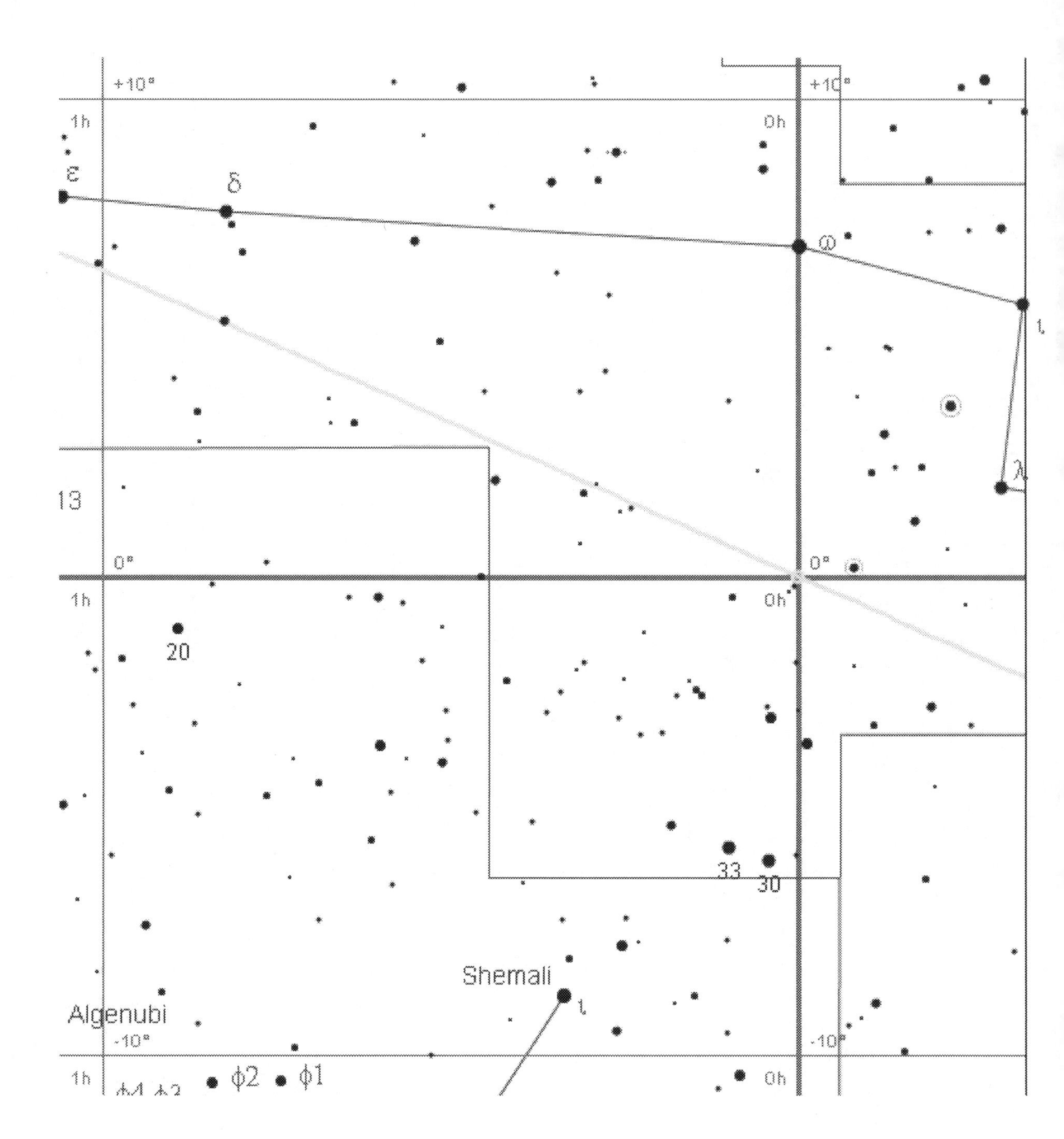
+10°
1h
0h
ε
δ
ω
ι
λ
13
0°
20
33
30
Shemali
ι
Algenubi
-10°
φ2
φ1

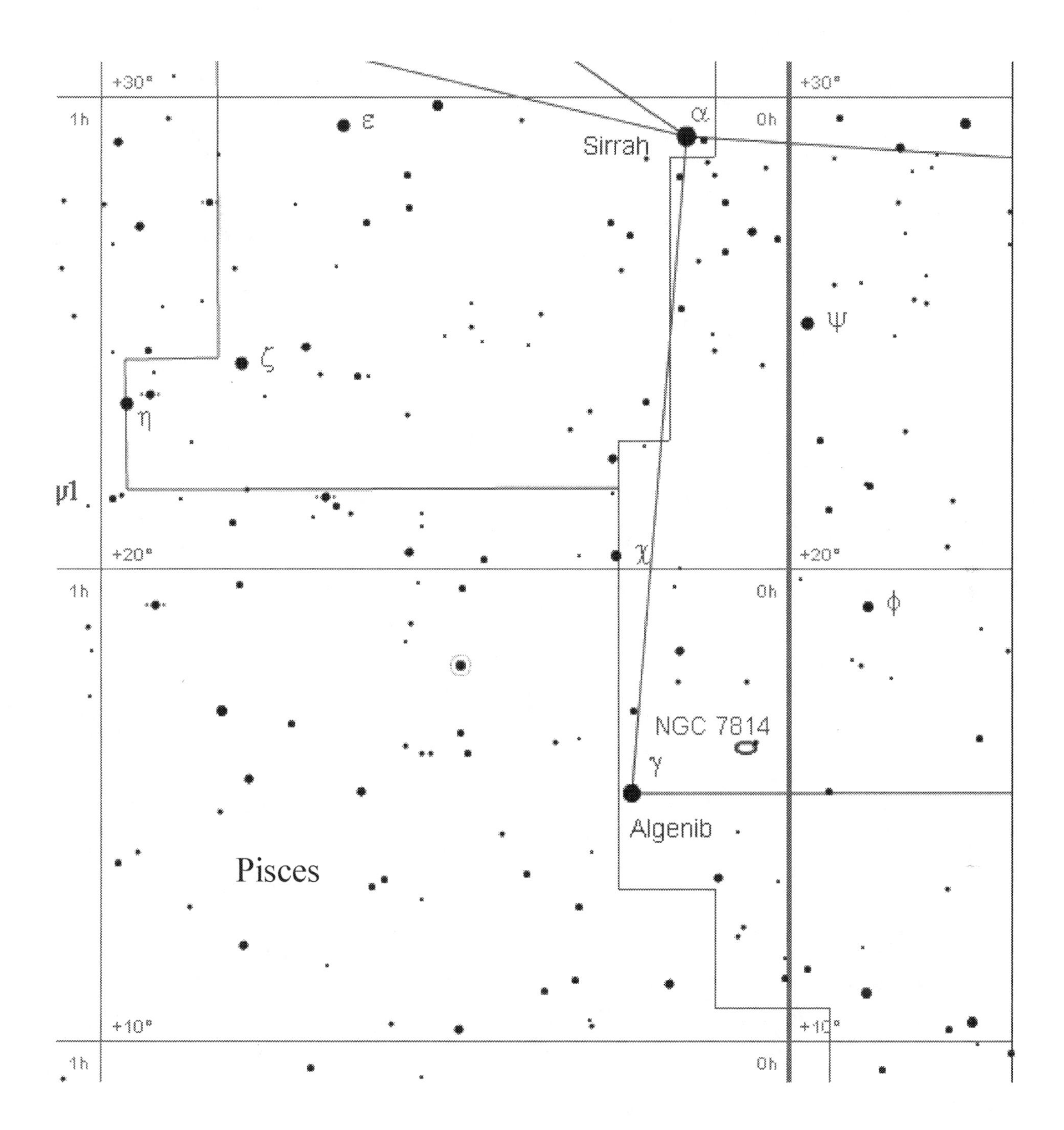

+30°
1h
ε
α
0h
Sirrah
ψ
ζ
η
μ1
+20°
χ
1h
0h
ϕ
NGC 7814
γ
Algenib
Pisces
+10°
1h
0h

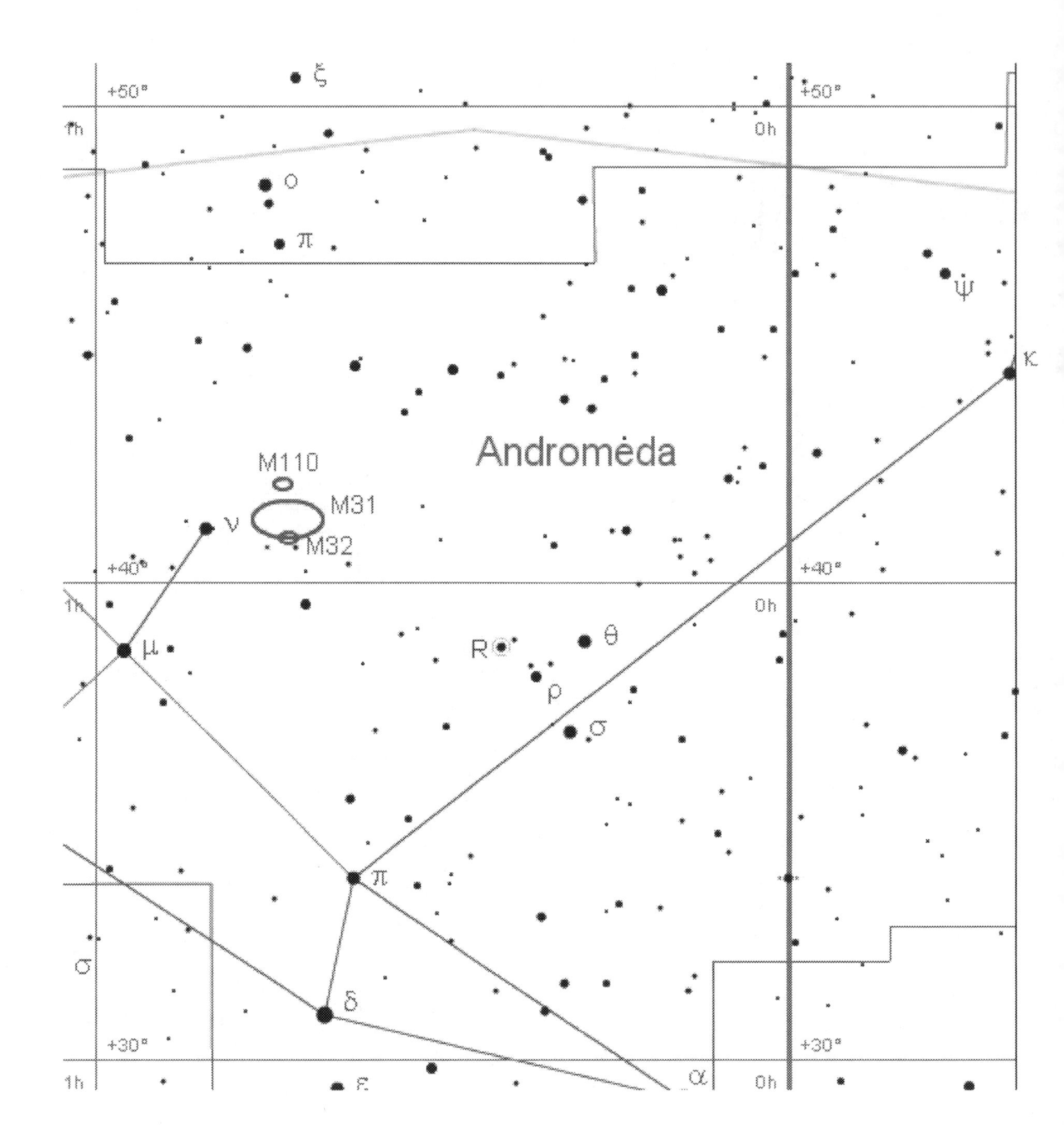
+50°
1h
0h
+50°
ξ
ο
π
ψ
κ
Andromeda
M110
M31
M32
ν
+40°
1h
0h
+40°
μ
R
θ
ρ
σ
π
σ
δ
+30°
1h
α
0h
+30°
ε

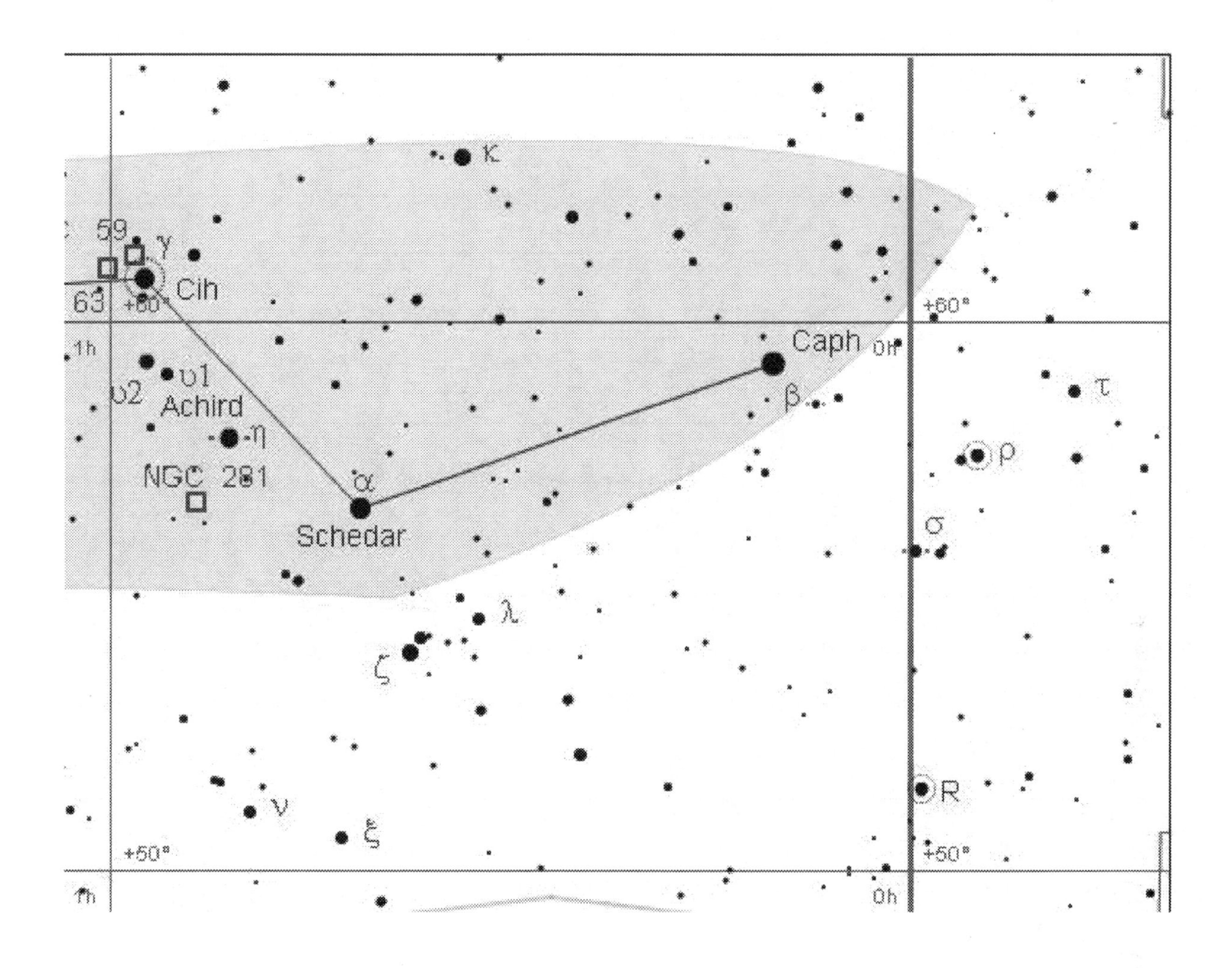
59
γ
63
Cih
+60°
1h
υ2
υ1
Achird
η
NGC 281
α
Schedar
κ
Caph
0h
β
+60°
τ
ρ
σ
λ
ζ
ν
ξ
R
+50°
+50°
1h
0h

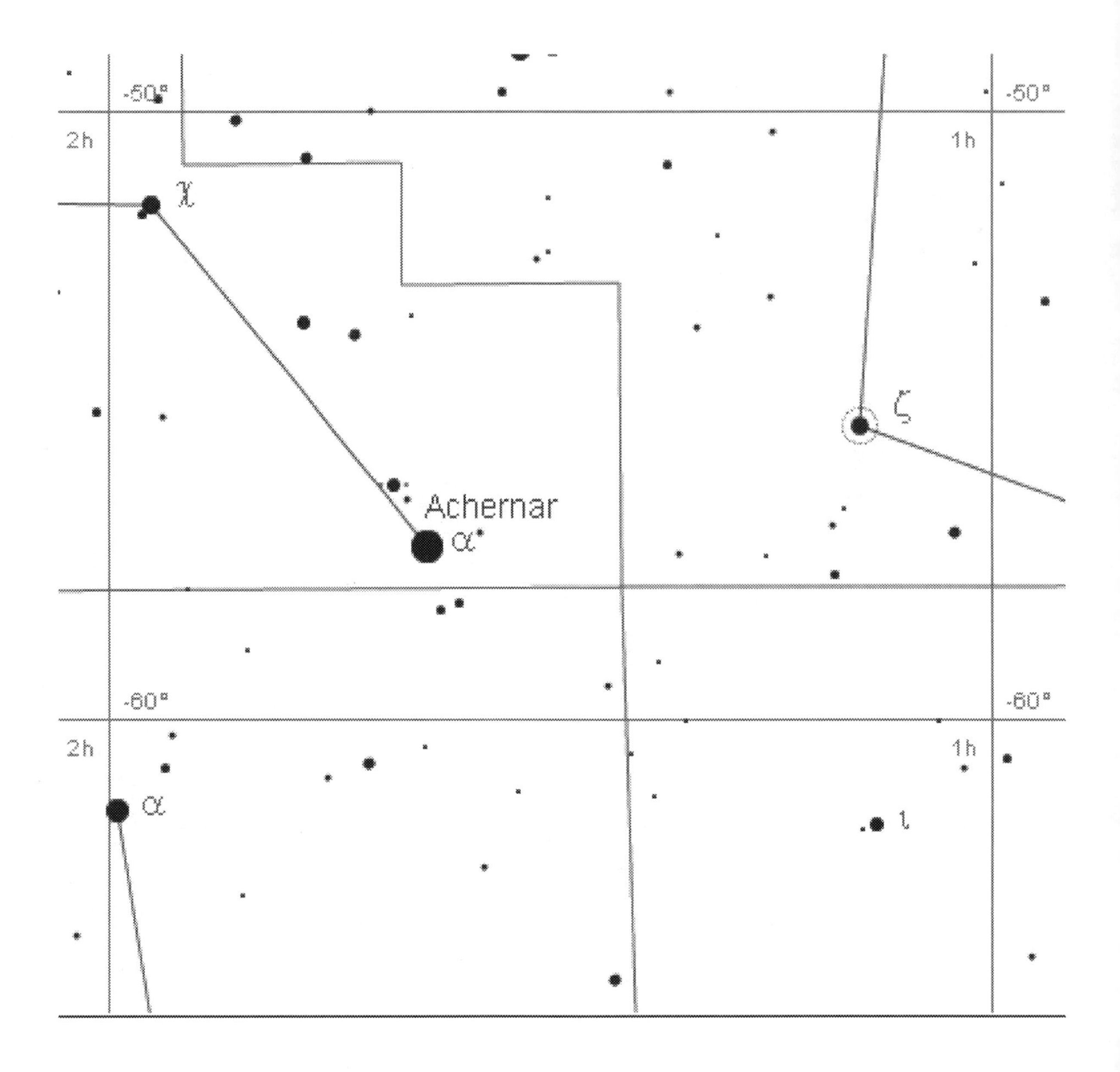
-50°
-50°
2h
1h
χ
ζ
Achernar
α*
-60°
-60°
2h
1h
α
ι

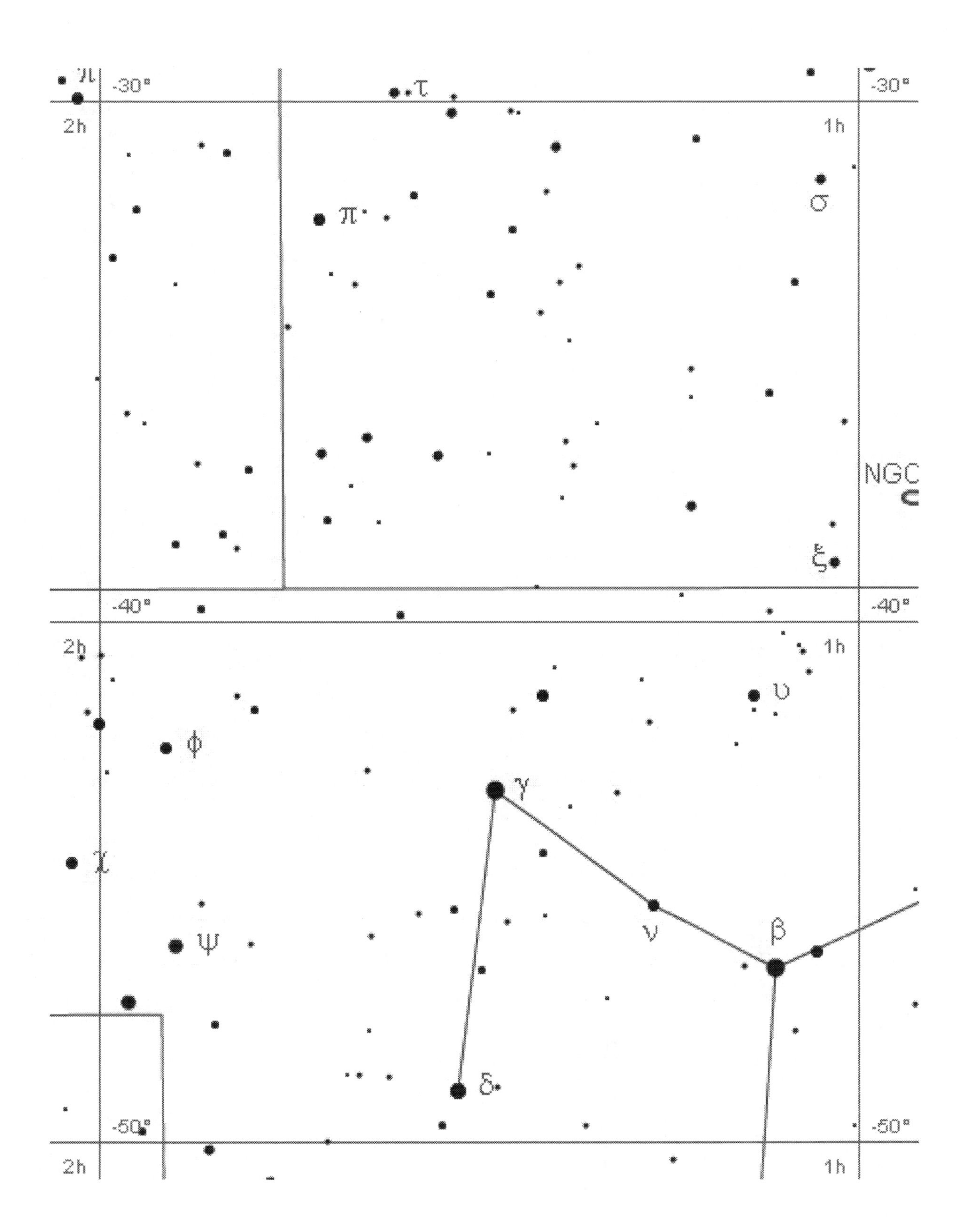
-30°
2h
1h
-30°
π
τ
σ
NGC
ξ
-40°
-40°
2h
1h
υ
φ
γ
χ
ν
β
ψ
δ
-50°
-50°
2h
1h

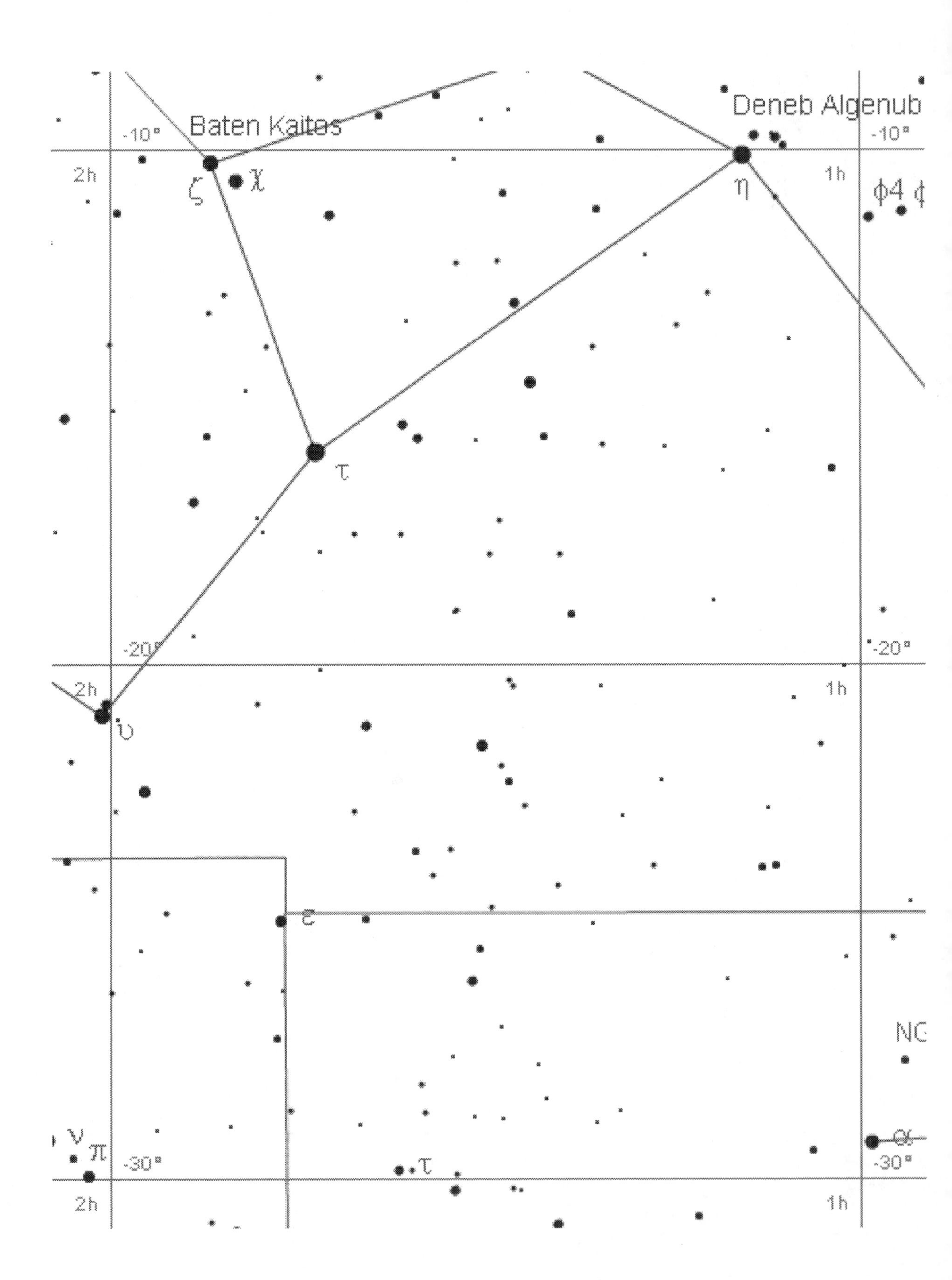

Baten Kaitos
Deneb Algenub
-10°
2h
1h
ζ
χ
η
φ4
τ
-20°
υ
ε
NG
ν
π
τ
α
-30°

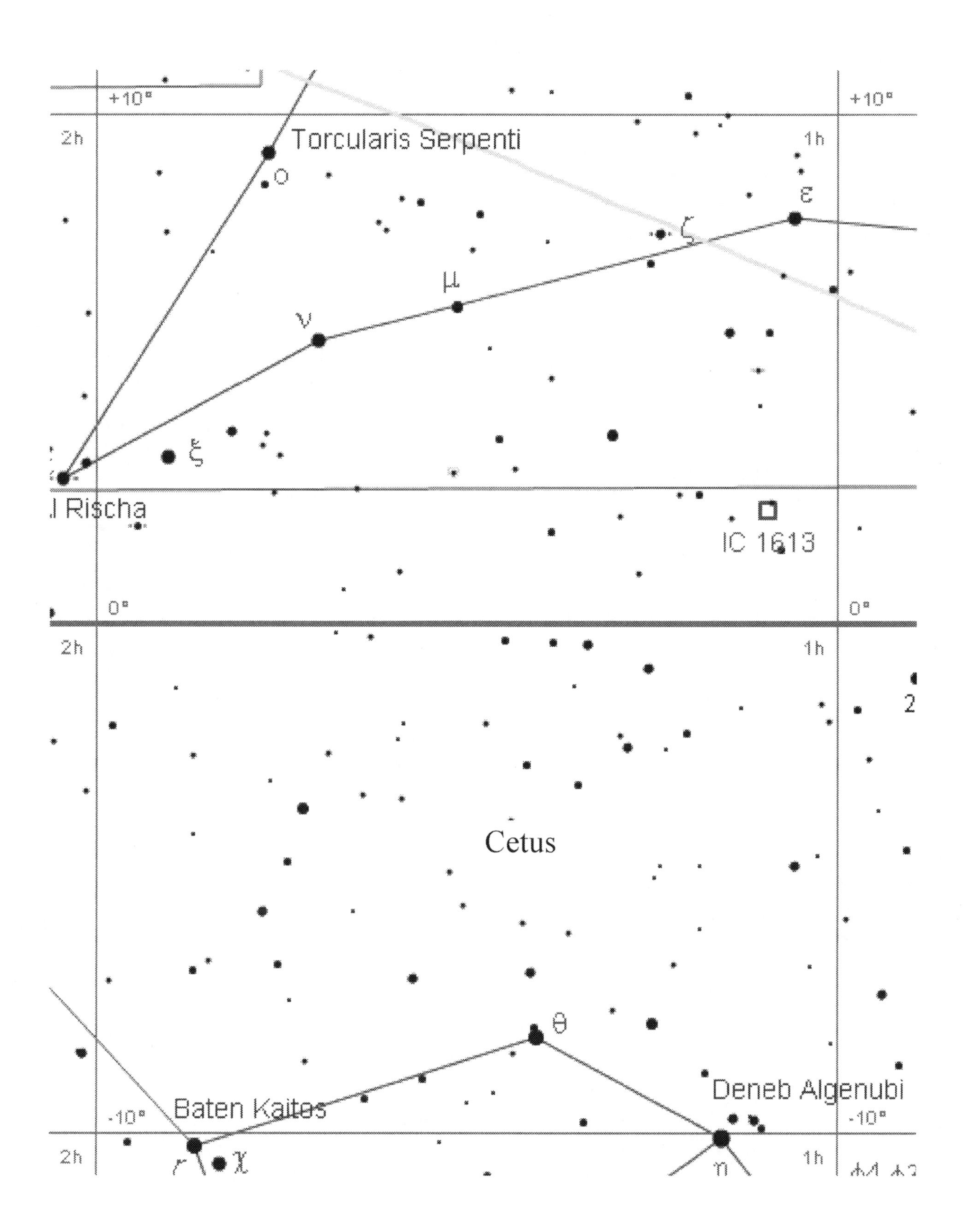

+10°
+10°
2h
1h
Torcularis Serpenti
ο
ε
ζ
μ
ν
ξ
Rischa
IC 1613
0°
0°
2h
1h
Cetus
θ
Deneb Algenubi
Baten Kaitos
-10°
-10°
2h
1h
χ
η

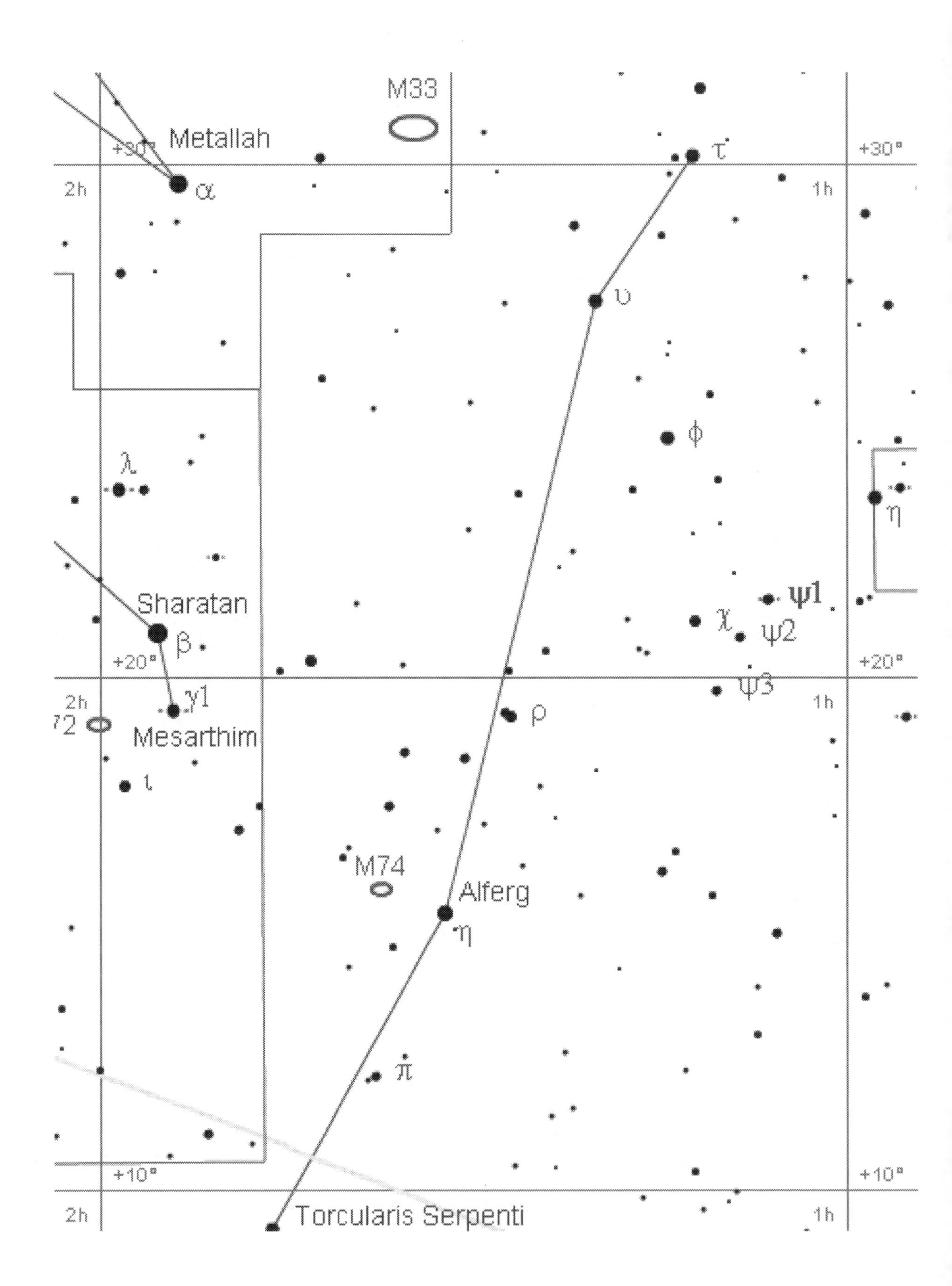
M33
Metallah
+30°
2h
α
1h
τ
υ
ϕ
λ
η
Sharatan
β
ψ1
χ
ψ2
+20°
ψ3
γ1
Mesarthim
ρ
ι
M74
Alferg
η
π
+10°
Torcularis Serpenti

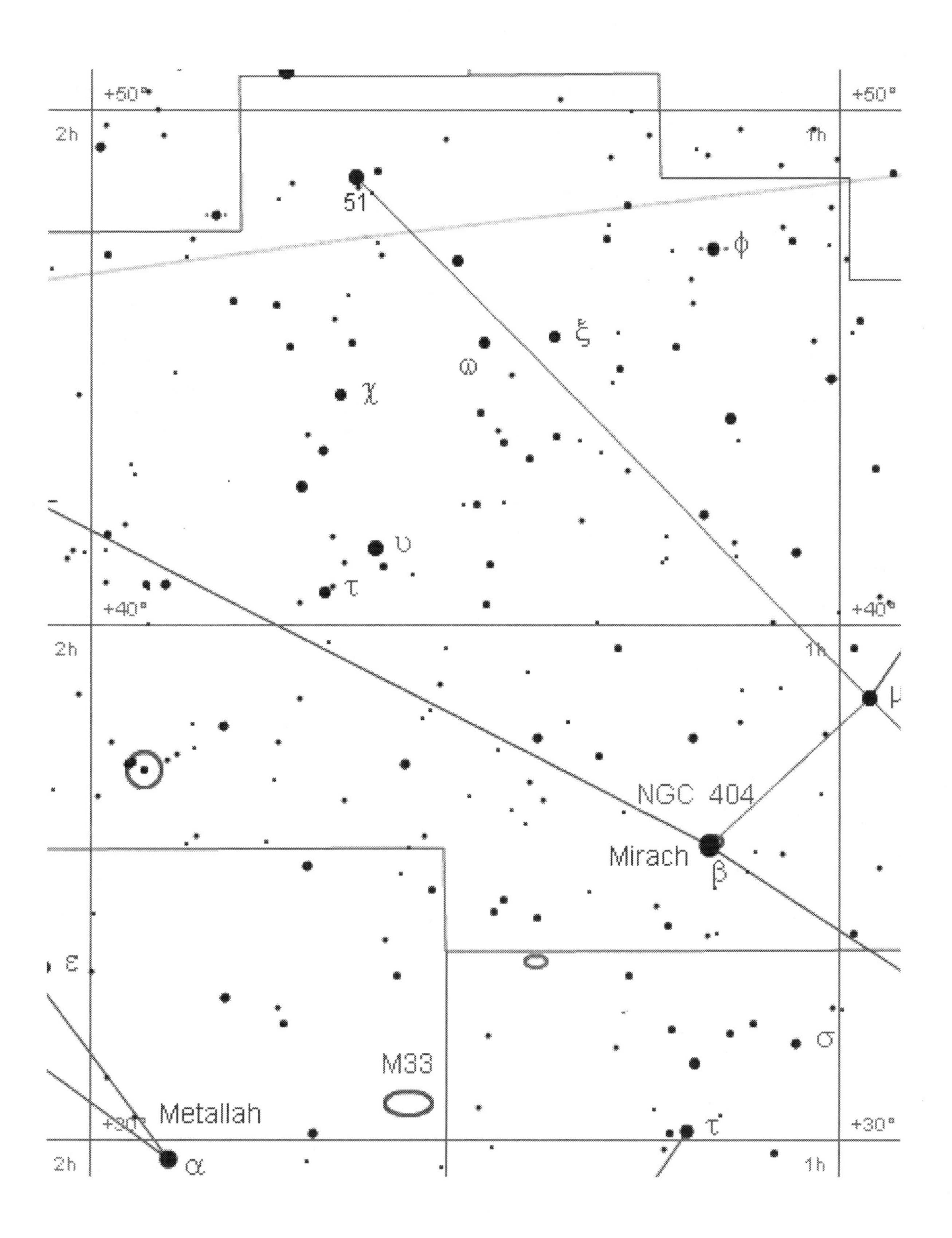

+50°
+50°
2h
1h
51
ϕ
ξ
ω
χ
υ
τ
+40°
+40°
2h
1h
μ
NGC 404
Mirach
β
ε
σ
M33
Metallah
+30°
+30°
τ
2h
α
1h

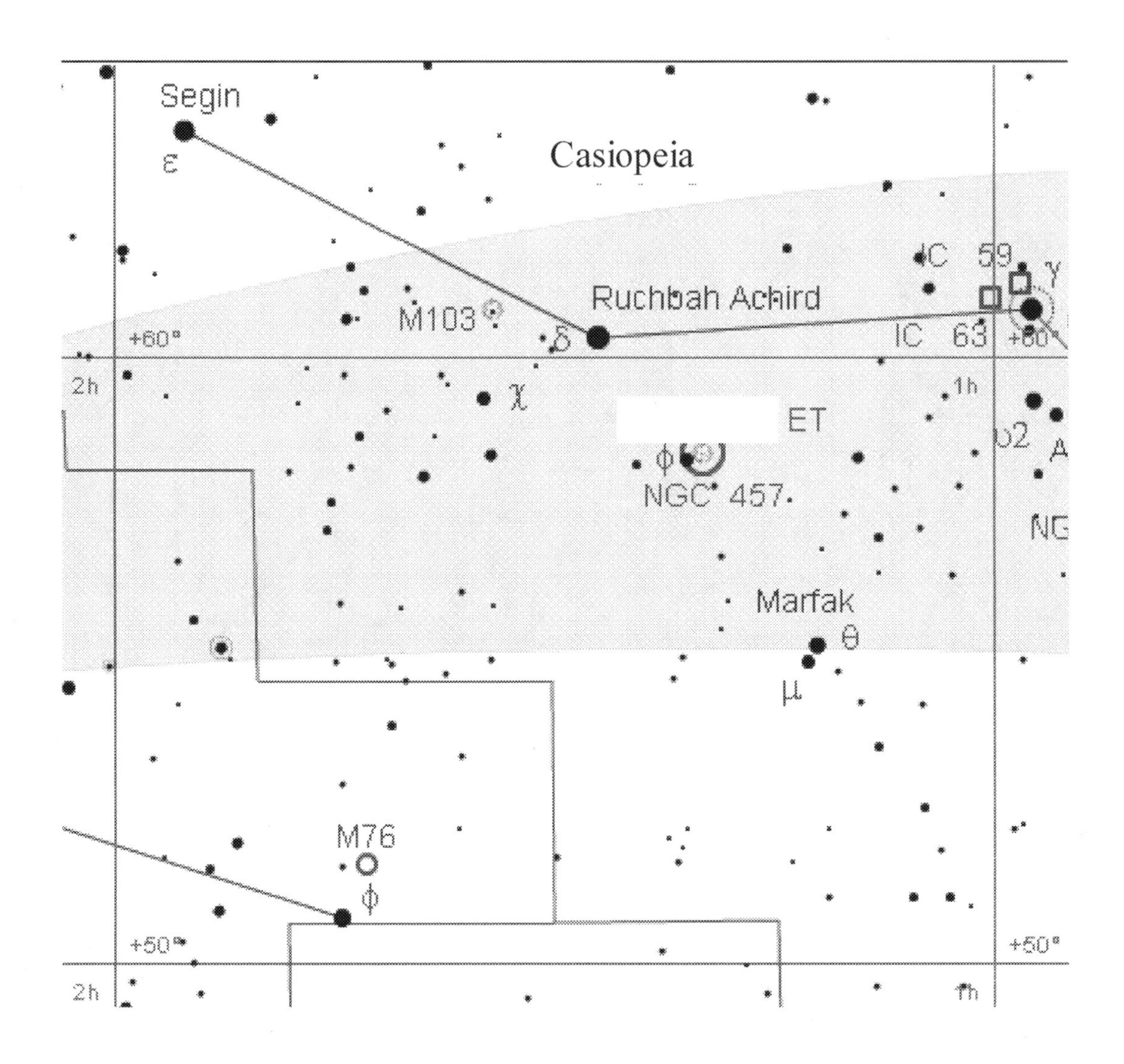
Segin
ε
Casiopeia
IC 59
γ
M103
Ruchbah Achird
δ
+60°
IC 63
+60°
2h
χ
1h
ET
φ
NGC 457
Marfak
θ
μ
M76
φ
+50°
+50°
2h
1h

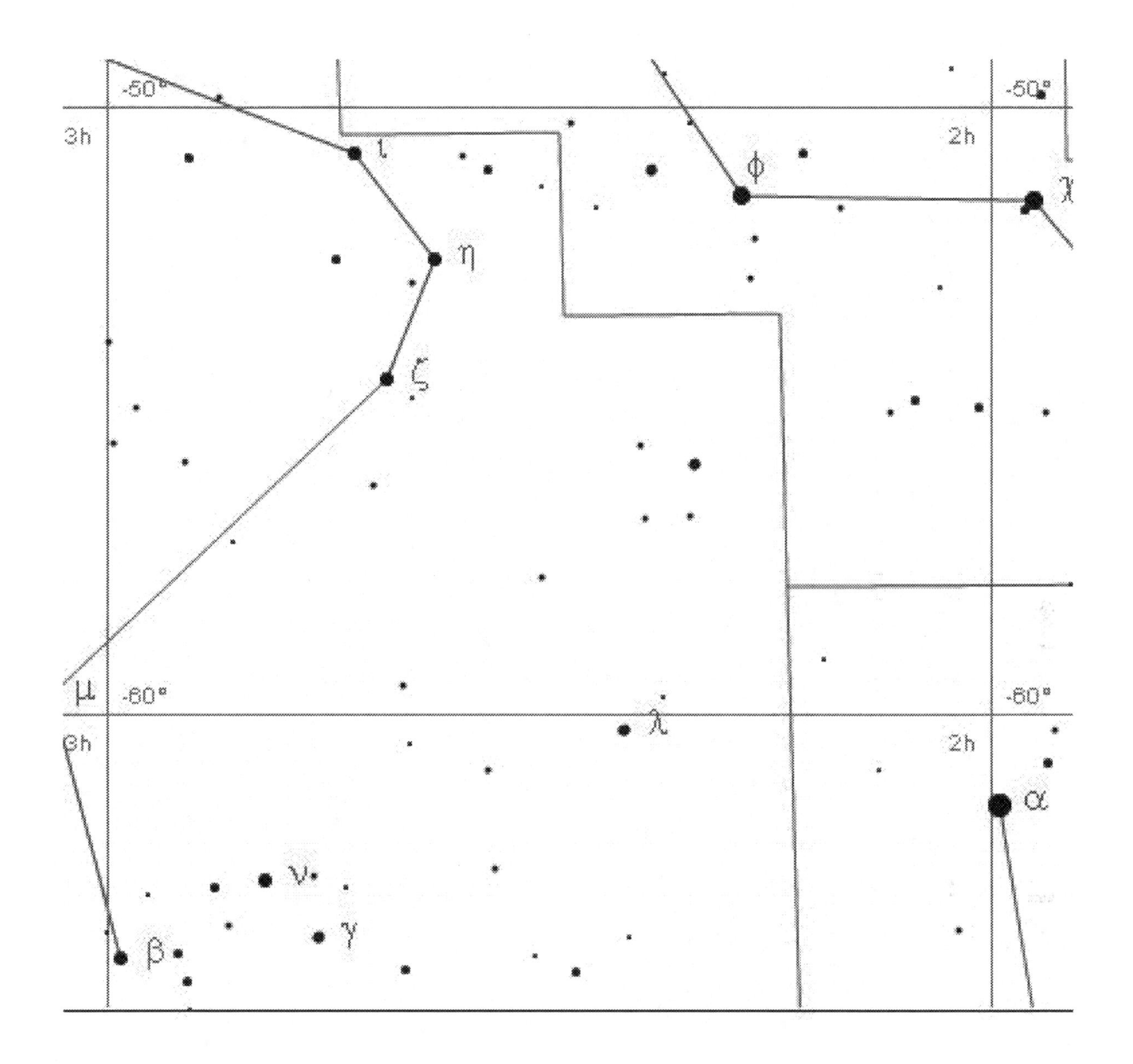
-50°
3h
2h
-50°
ι
φ
η
ζ
μ
-60°
-60°
3h
λ
2h
α
ν
γ
β

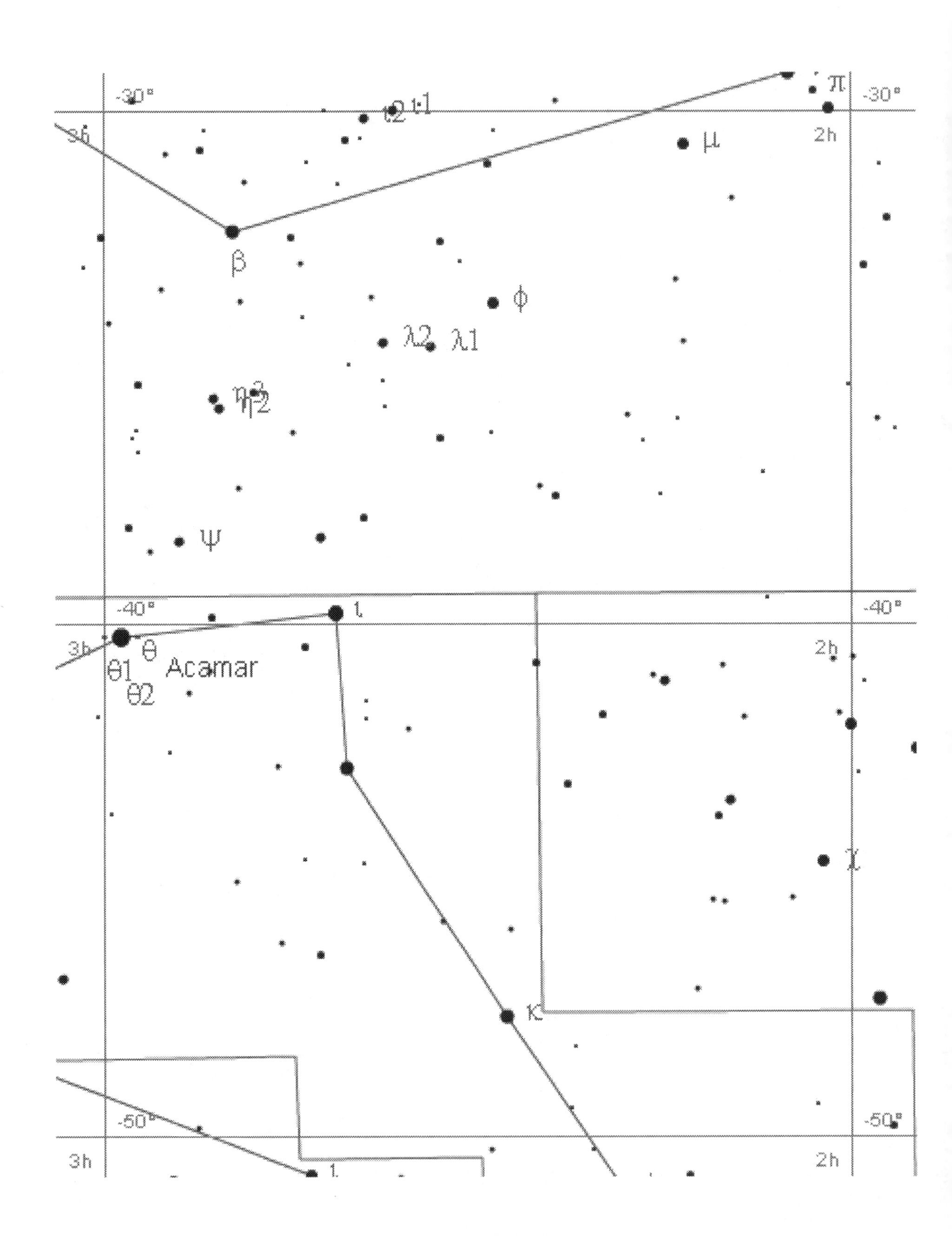
-30°
3h
ι2 ι1
π
-30°
2h
μ
β
φ
λ2 λ1
η3
η2
ψ
-40°
ι
-40°
3h
θ
θ1
Acamar
θ2
2h
χ
κ
-50°
-50°
3h
2h

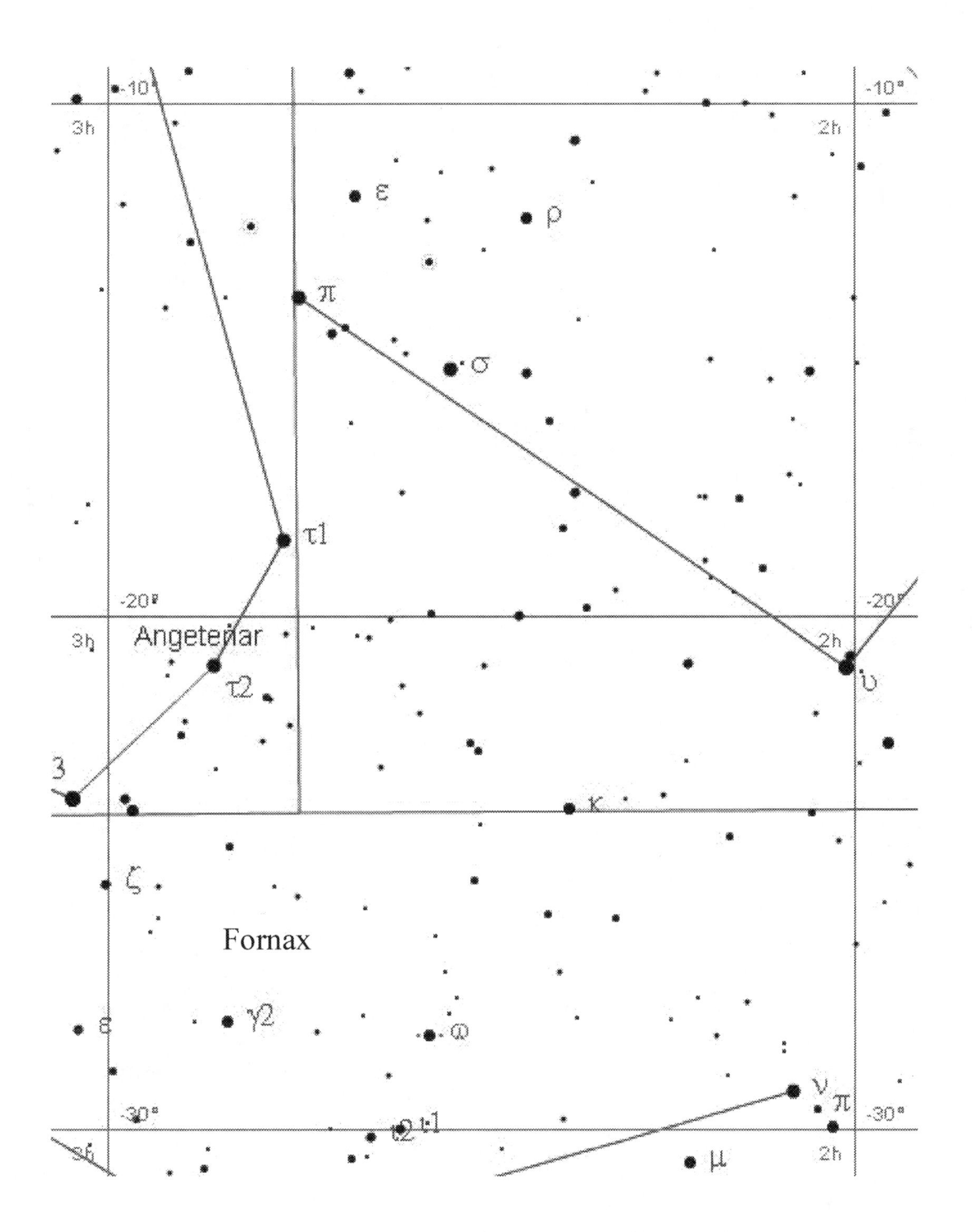
-10°
3h
2h
-10°
ε
ρ
π
σ
τ1
-20°
-20°
3h
Angetenar
2h
τ2
υ
κ
ζ
Fornax
ε
γ2
ω
ν
π
-30°
-30°
μ
2h

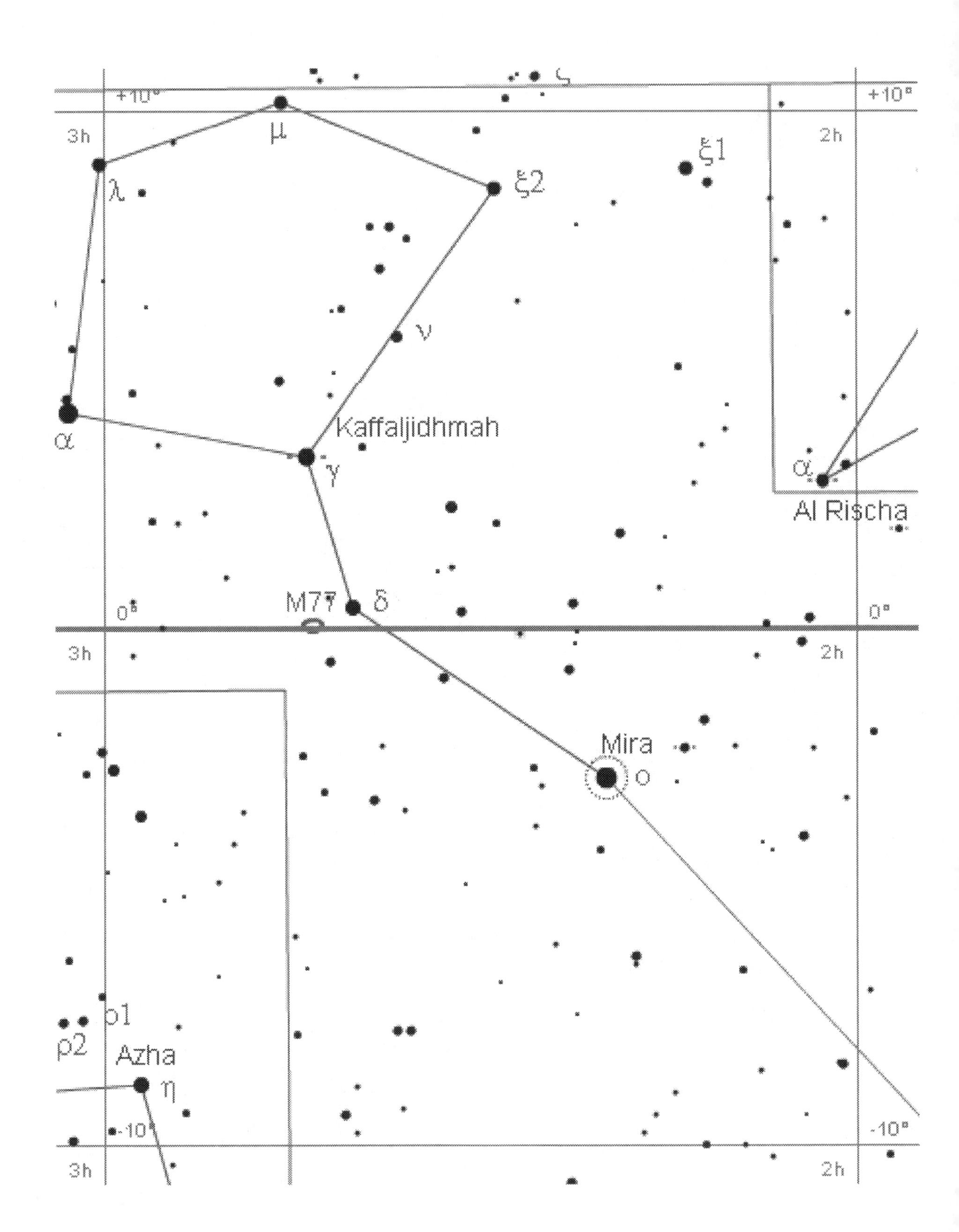
+10°
3h
μ
λ
ξ1
ξ2
2h
ν
Kaffaljidhmah
α
γ
α
Al Rischa
M77
δ
0°
3h
2h
Mira
o
ο1
ρ2
Azha
η
-10°
3h
2h

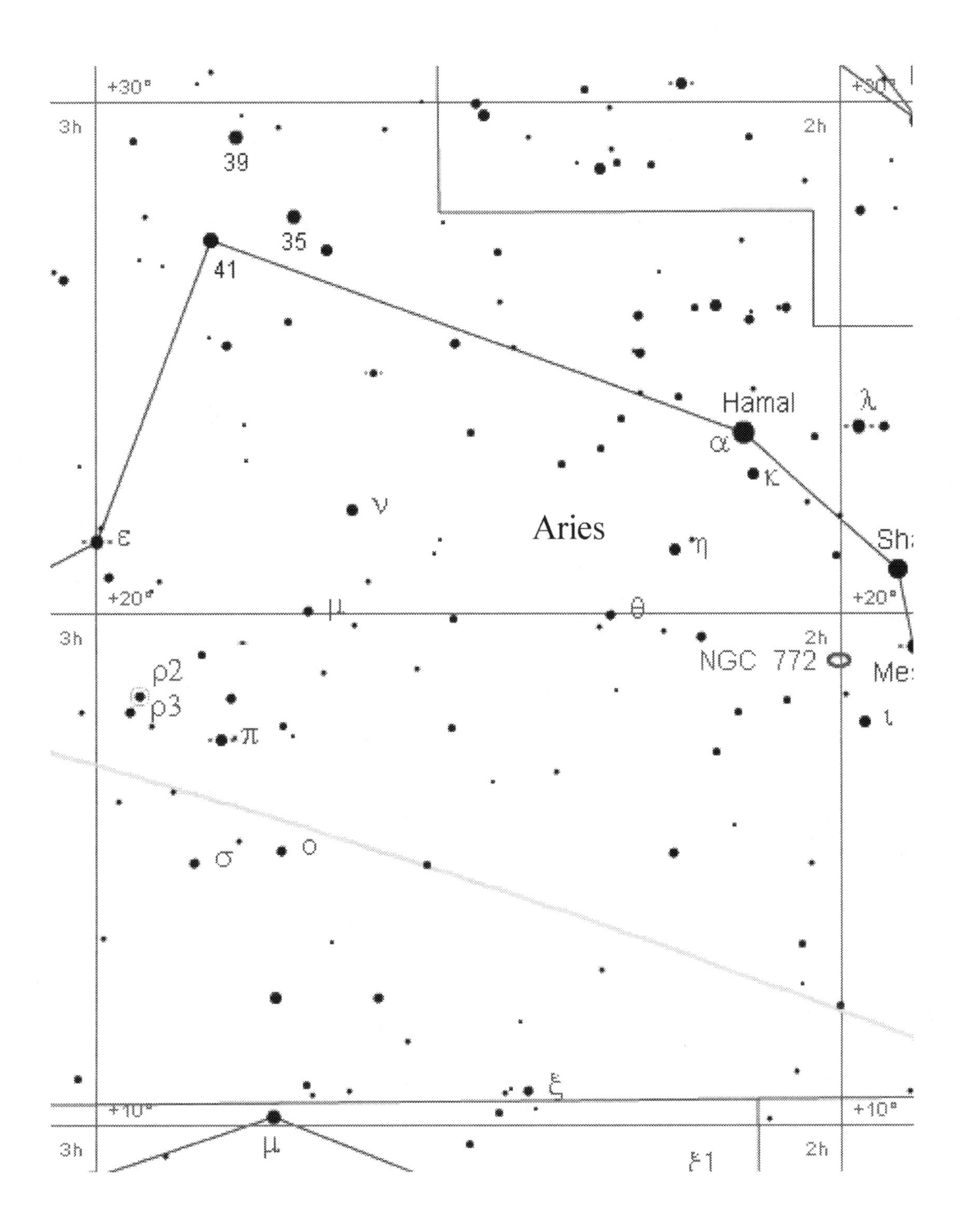
+30°
3h
39
35
41
2h
Hamal
α
κ
λ
ν
Aries
ε
η
Sh
+20°
μ
θ
NGC 772
Me
ρ2
ρ3
π
ι
σ
o
ξ
+10°
μ
ξ1

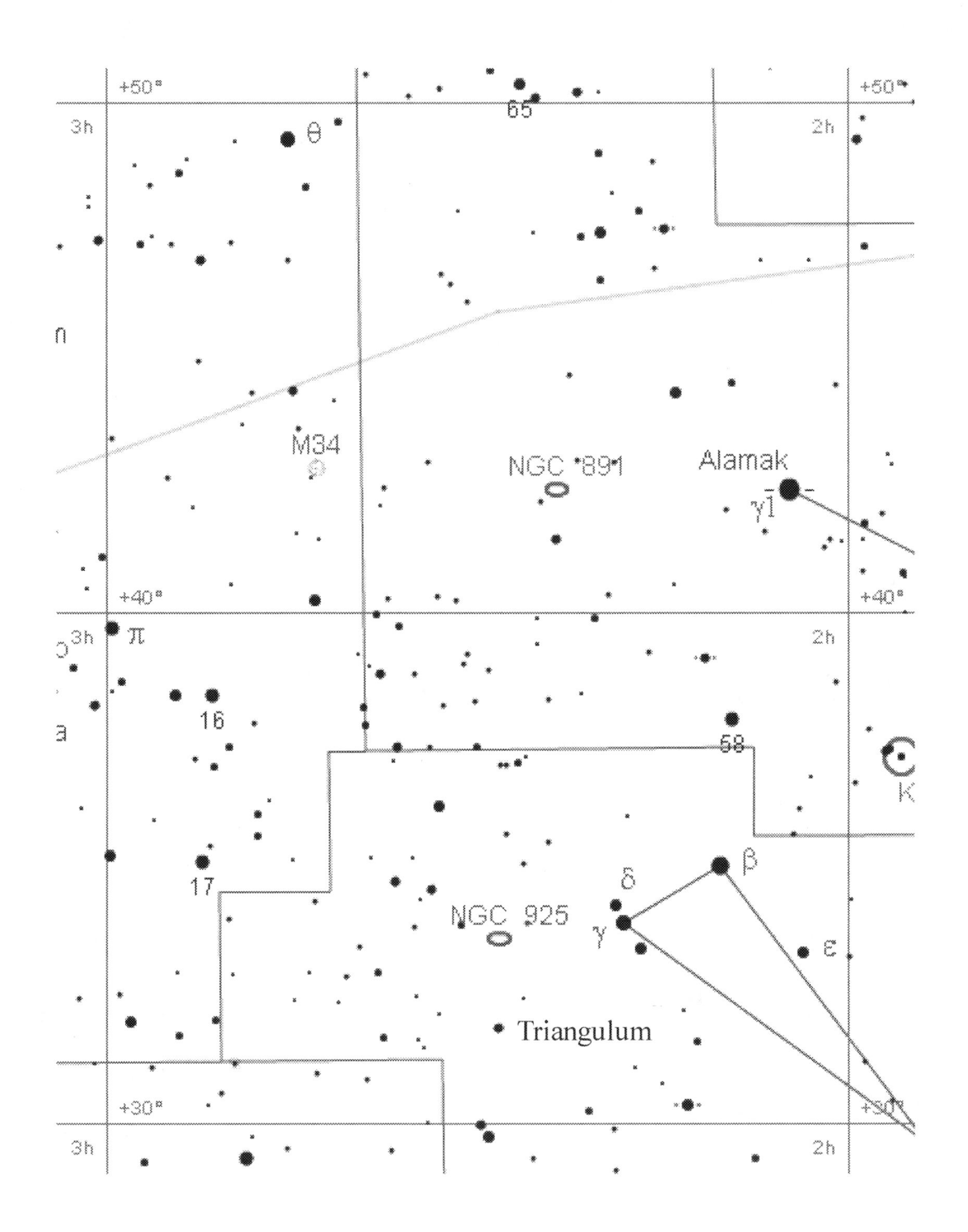

+50°
65
3h
θ
2h
+50°
M34
NGC 891
Alamak
γ1
+40°
+40°
3h
π
2h
16
58
K
17
δ
β
NGC 925
γ
ε
Triangulum
+30°
+30°
3h
2h

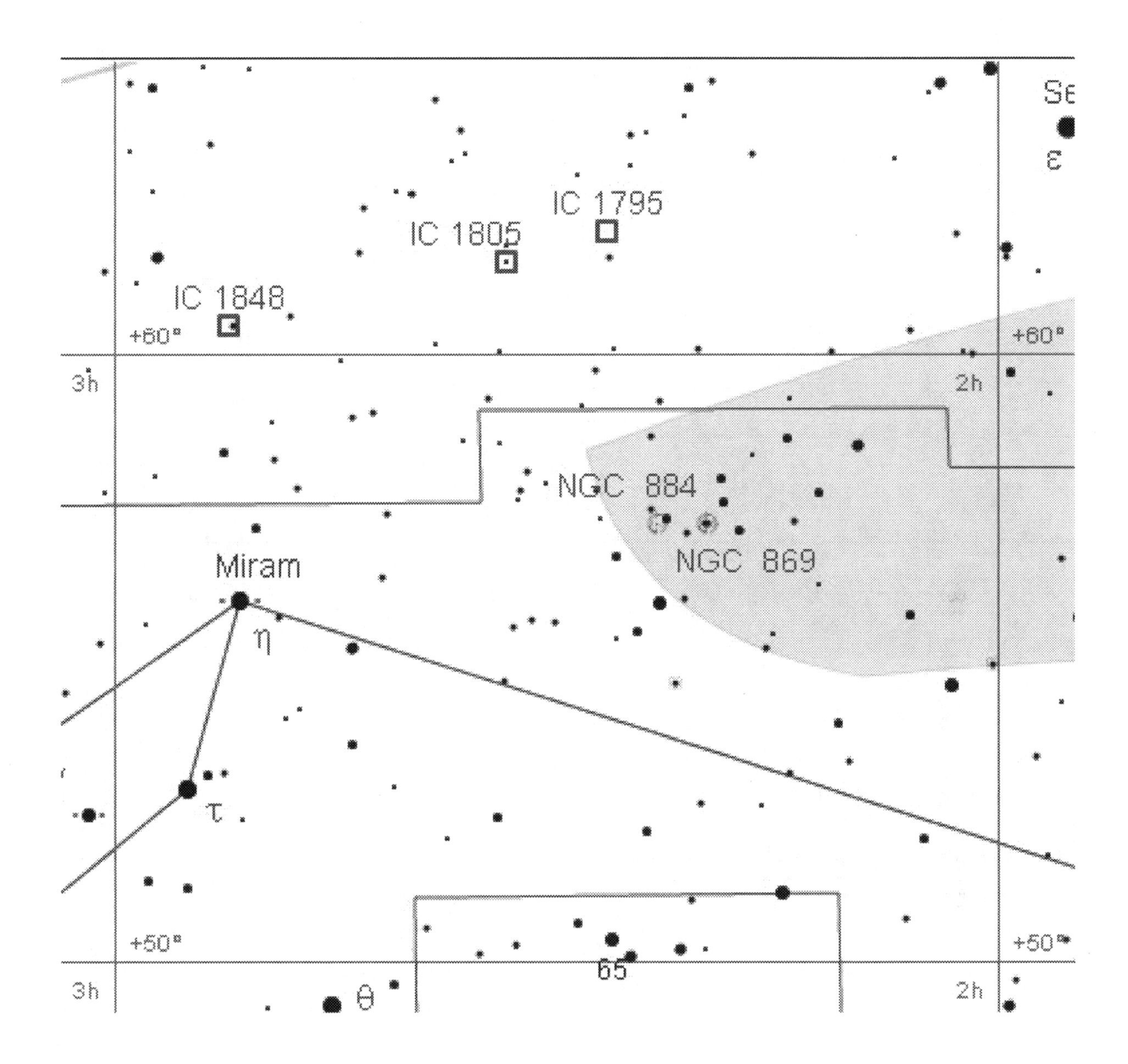
IC 1795
IC 1805
IC 1848
+60°
3h
2h
NGC 884
NGC 869
Miram
η
τ
+50°
65
θ
ε

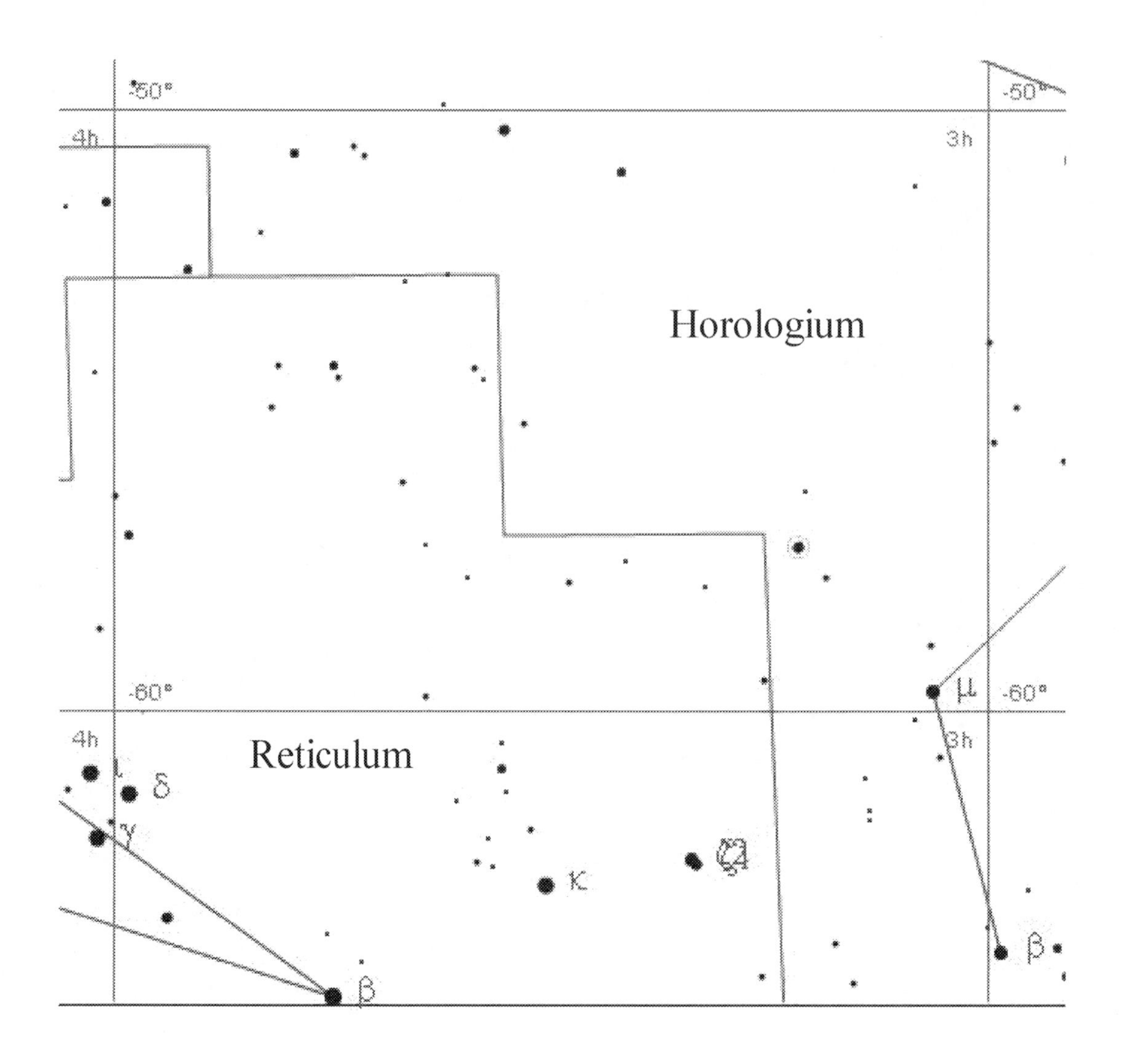

-50°
-50°
4h
3h
Horologium
-60°
μ
-60°
4h
3h
Reticulum
ι
δ
γ
κ
ζ
β
β

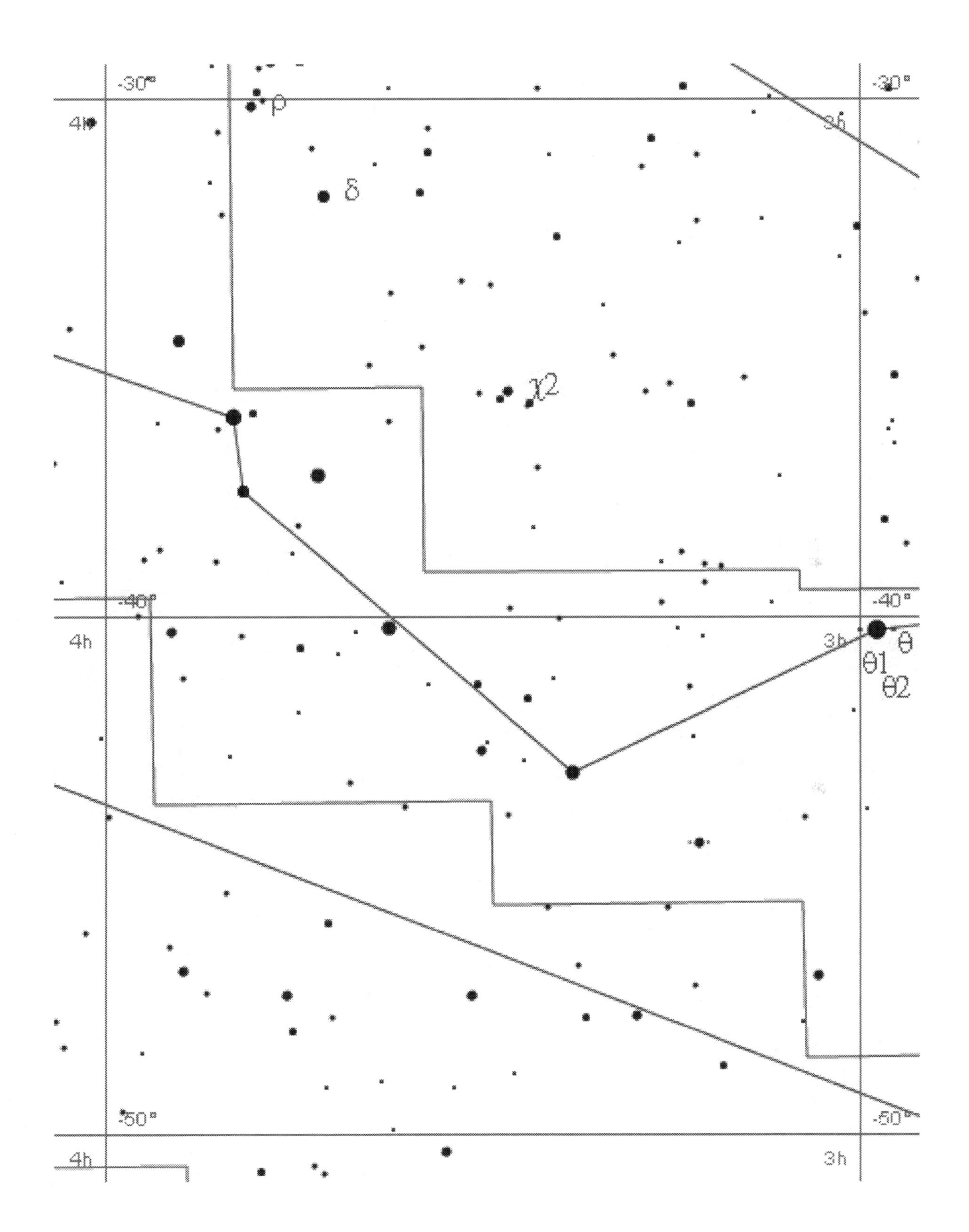
-30°
4h
ρ
δ
3h
χ2
-40°
4h
3h
θ
θ1
θ2
-50°
4h
3h

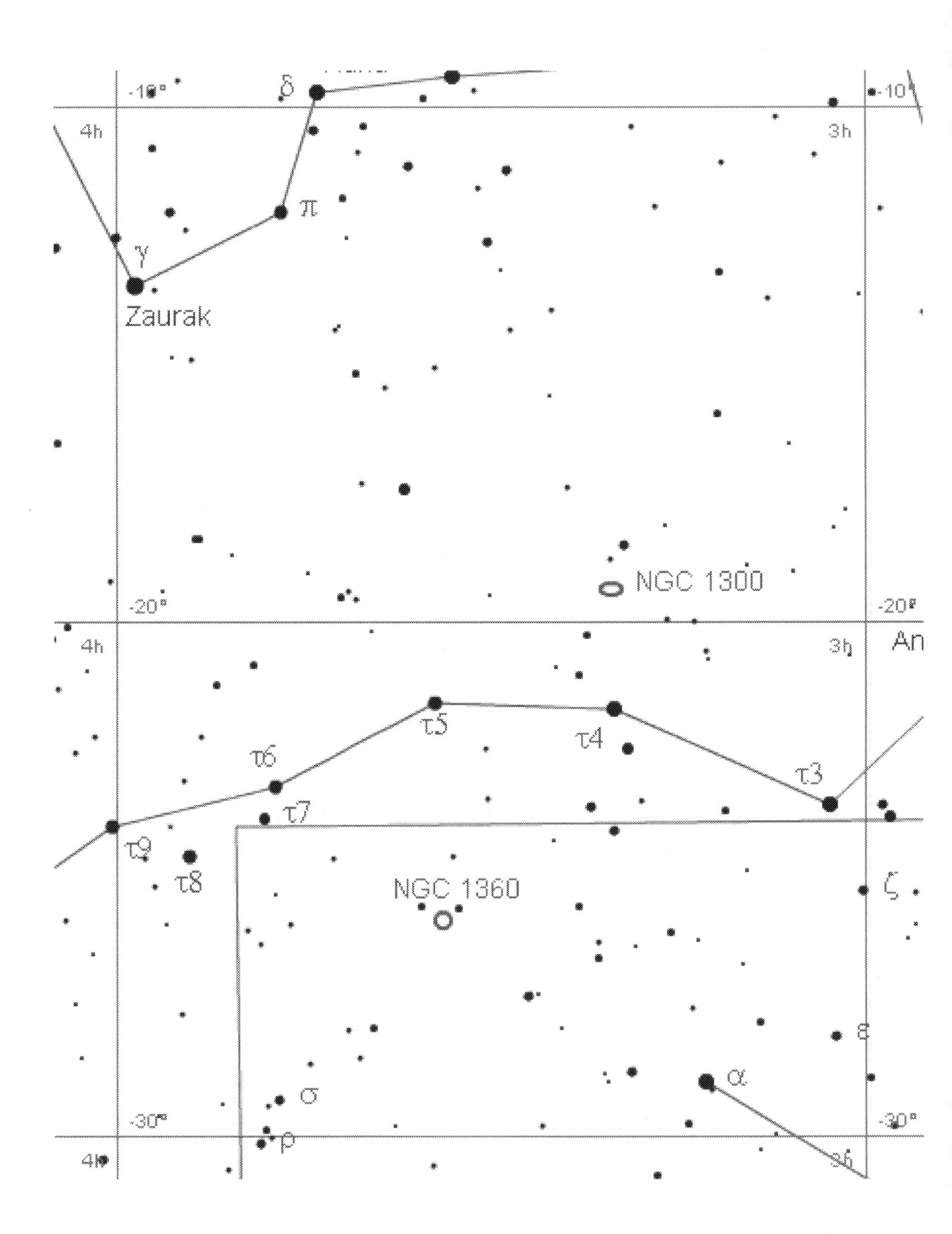

δ
-10°
4h
3h
π
γ
Zaurak
NGC 1300
-20°
An
τ5
τ4
τ6
τ3
τ7
τ9
τ8
NGC 1360
ζ
σ
α
-30°
ρ

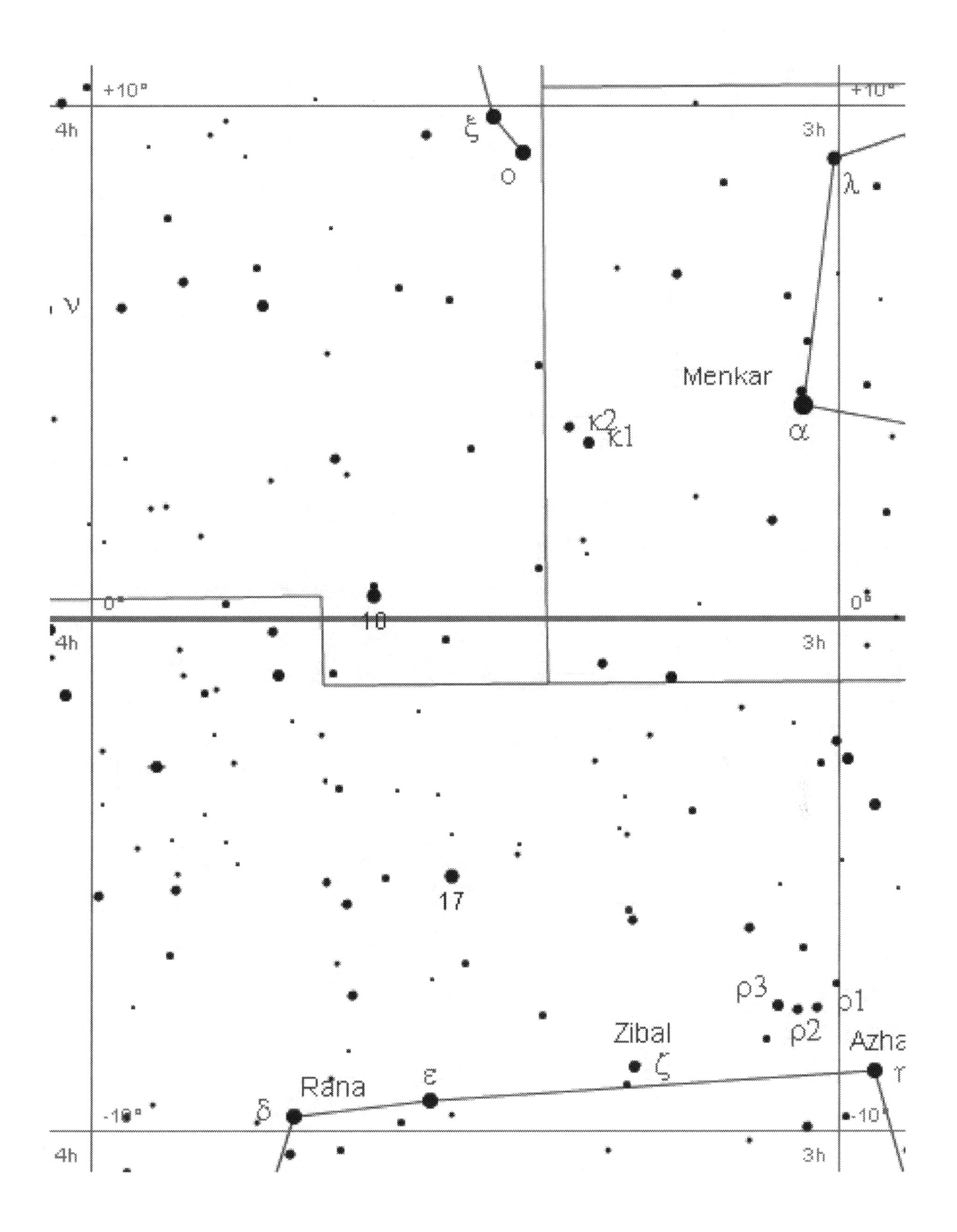

+10°
4h
ξ
o
3h
λ
ν
Menkar
κ2
κ1
α
0°
10
4h
3h
17
ρ3
ρ1
ρ2
Zibal
ζ
Azha
Rana
ε
δ
-10°
4h
3h

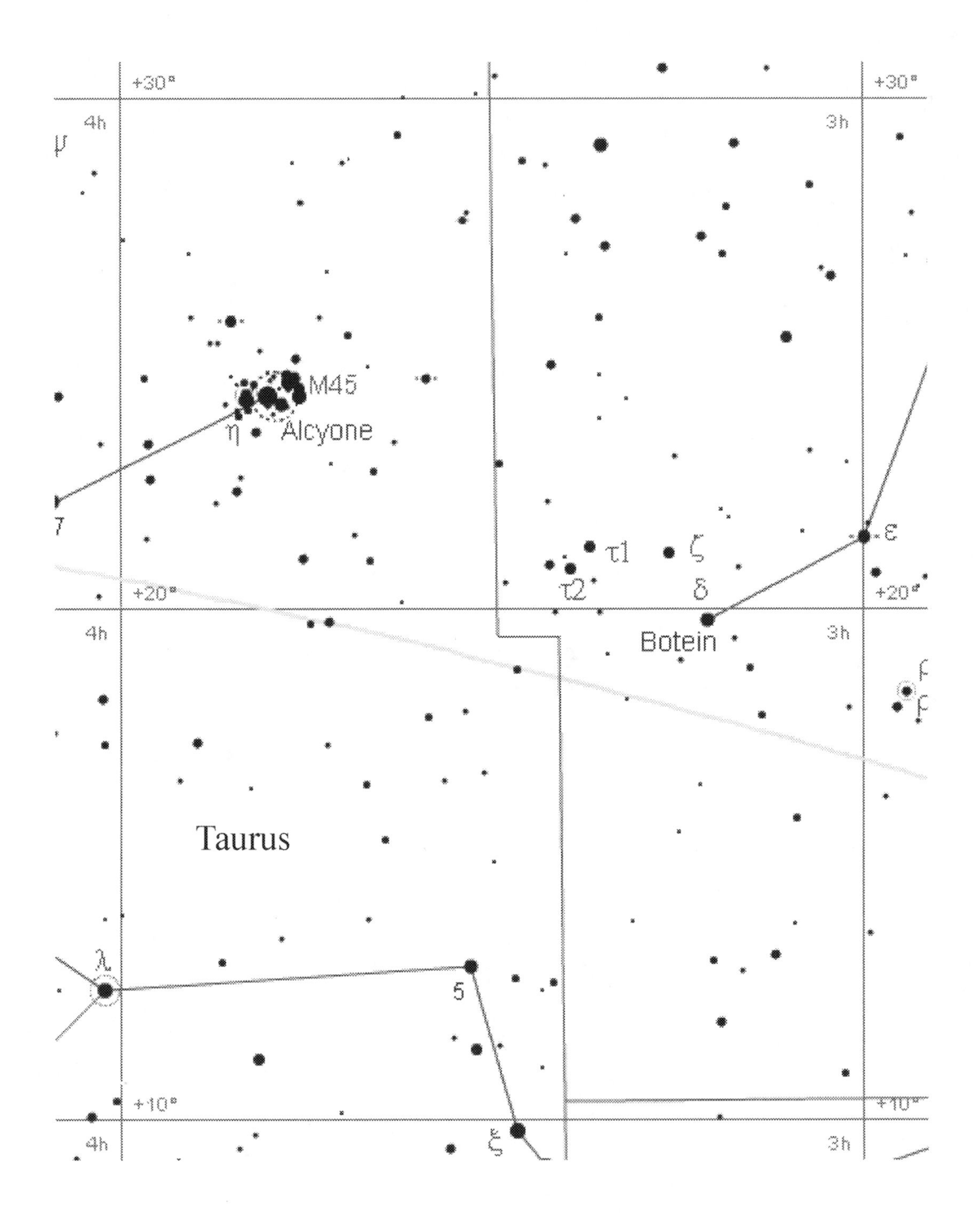
+30°
4h
3h
M45
Alcyone
η
τ1
ζ
τ2
δ
ε
+20°
Botein
Taurus
λ
5
+10°
ξ

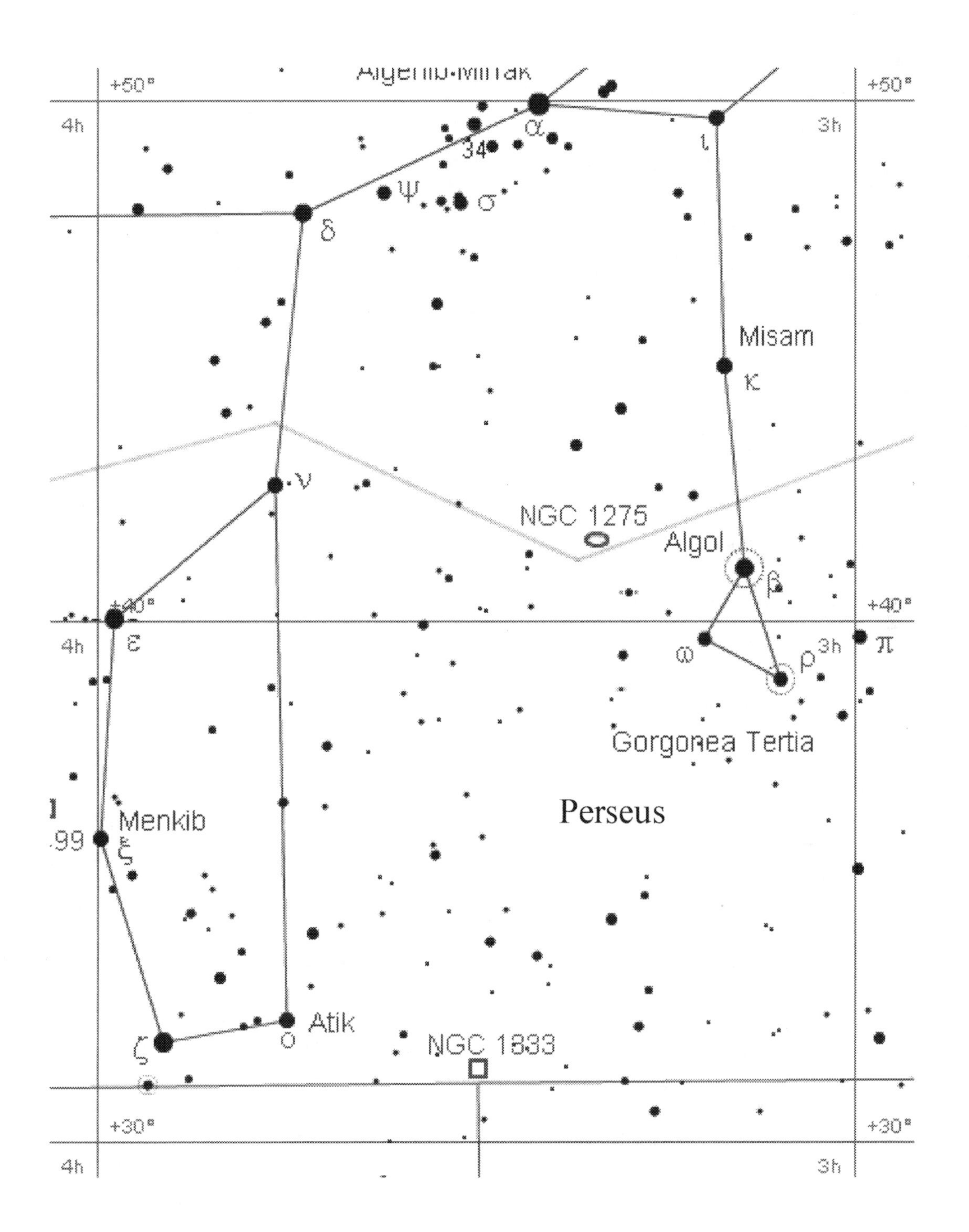
+50°
4h
3h
α
34
ψ
σ
δ
ι
Misam
κ
ν
NGC 1275
Algol
β
+40°
ε
ω
ρ
π
Gorgonea Tertia
Perseus
Menkib
.99
ξ
Atik
o
ζ
NGC 1333
+30°

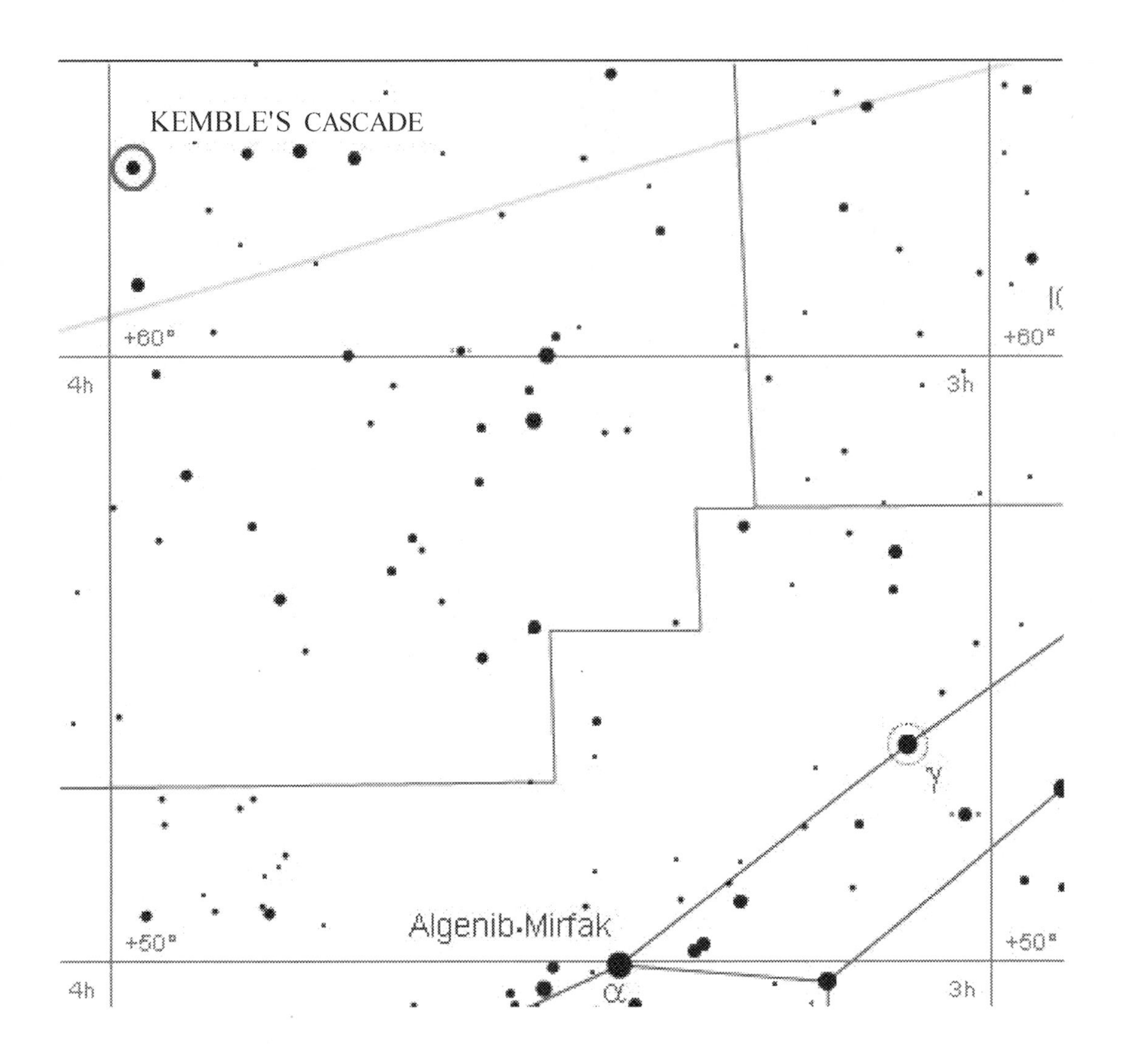

KEMBLE'S CASCADE
+60°
4h
3h
+60°
γ
Algenib-Mirfak
+50°
4h
α
3h
+50°

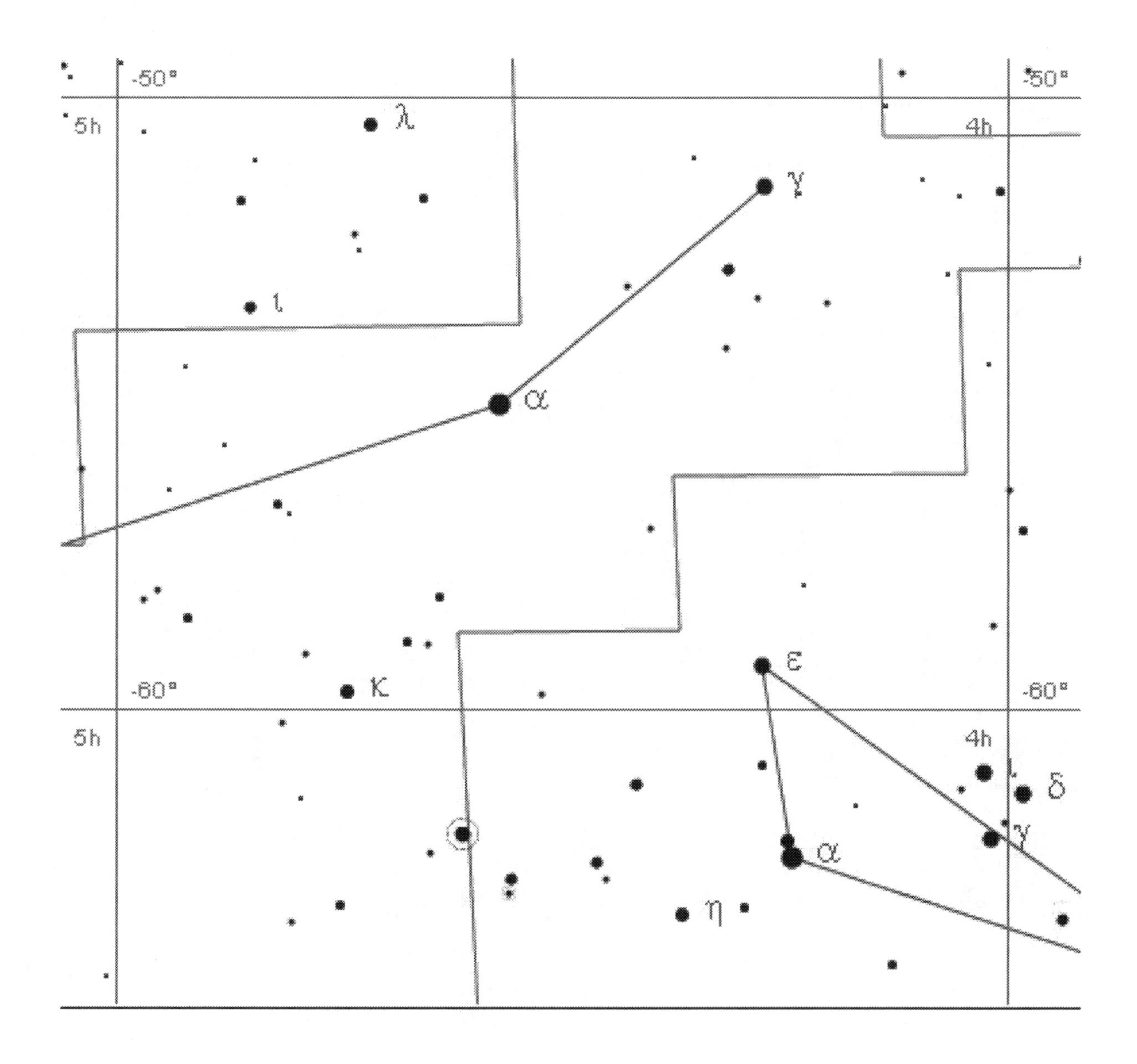
-50°
5h
λ
γ
ι
α
4h
-60°
κ
ε
δ
η

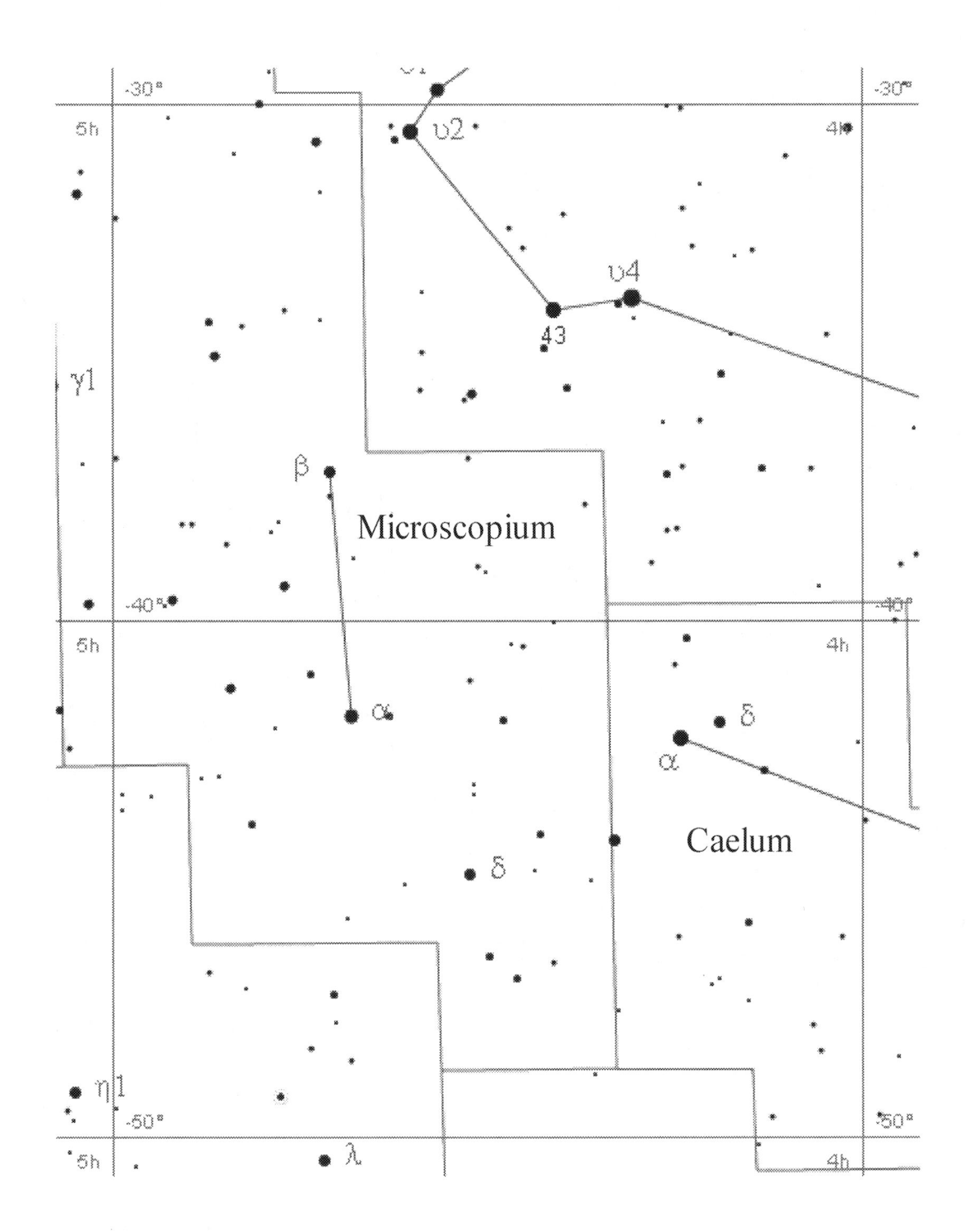
-30°
5h
υ2
4h
υ4
43
γ1
β
Microscopium
-40°
5h
4h
α
δ
α
Caelum
δ
η1
-50°
5h
λ
4h
-50°
-40°
-30°

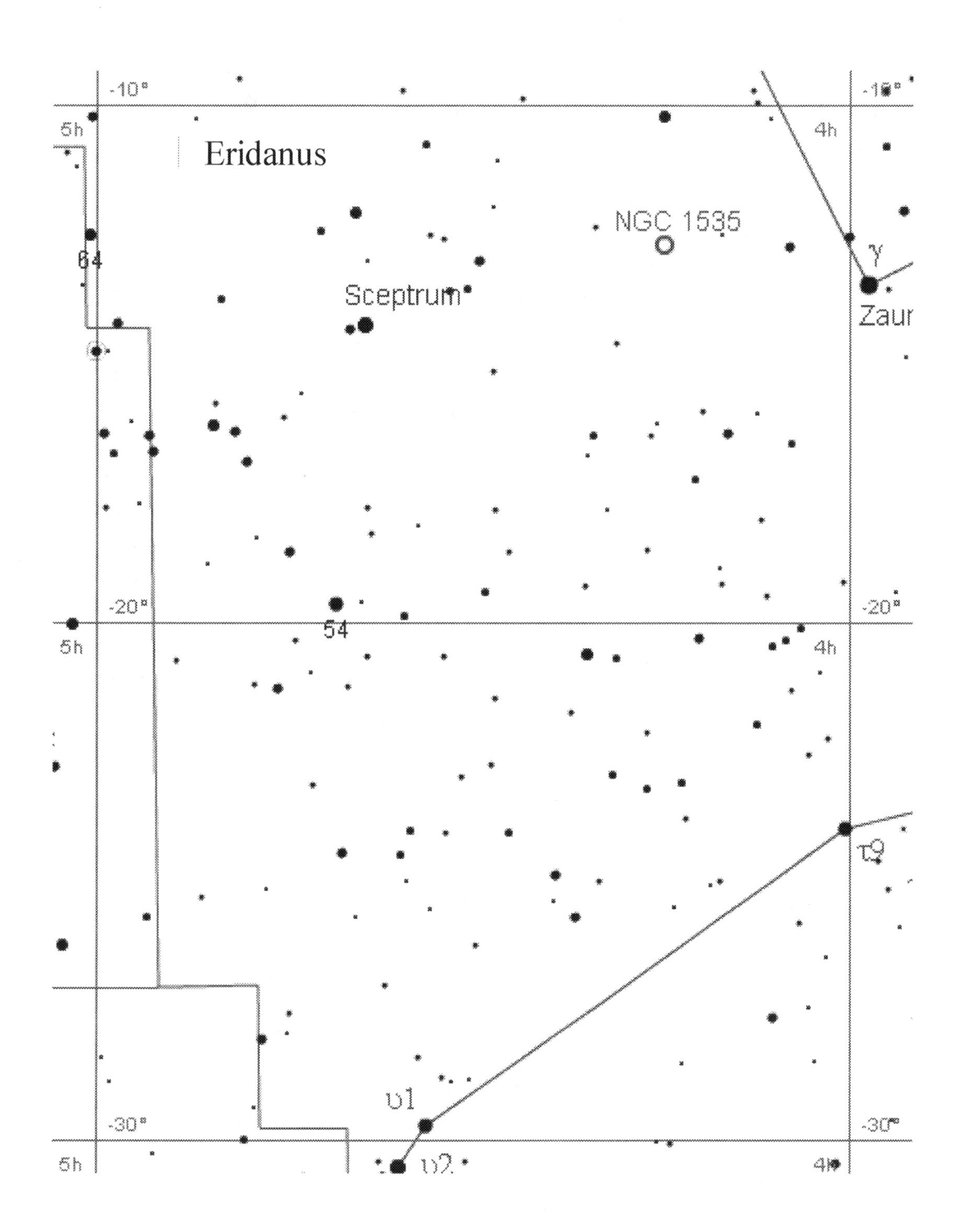

Eridanus
NGC 1535
Sceptrum
Zaur
64
54
γ
τ9
υ1
υ2
-10°
-20°
-30°
5h
4h

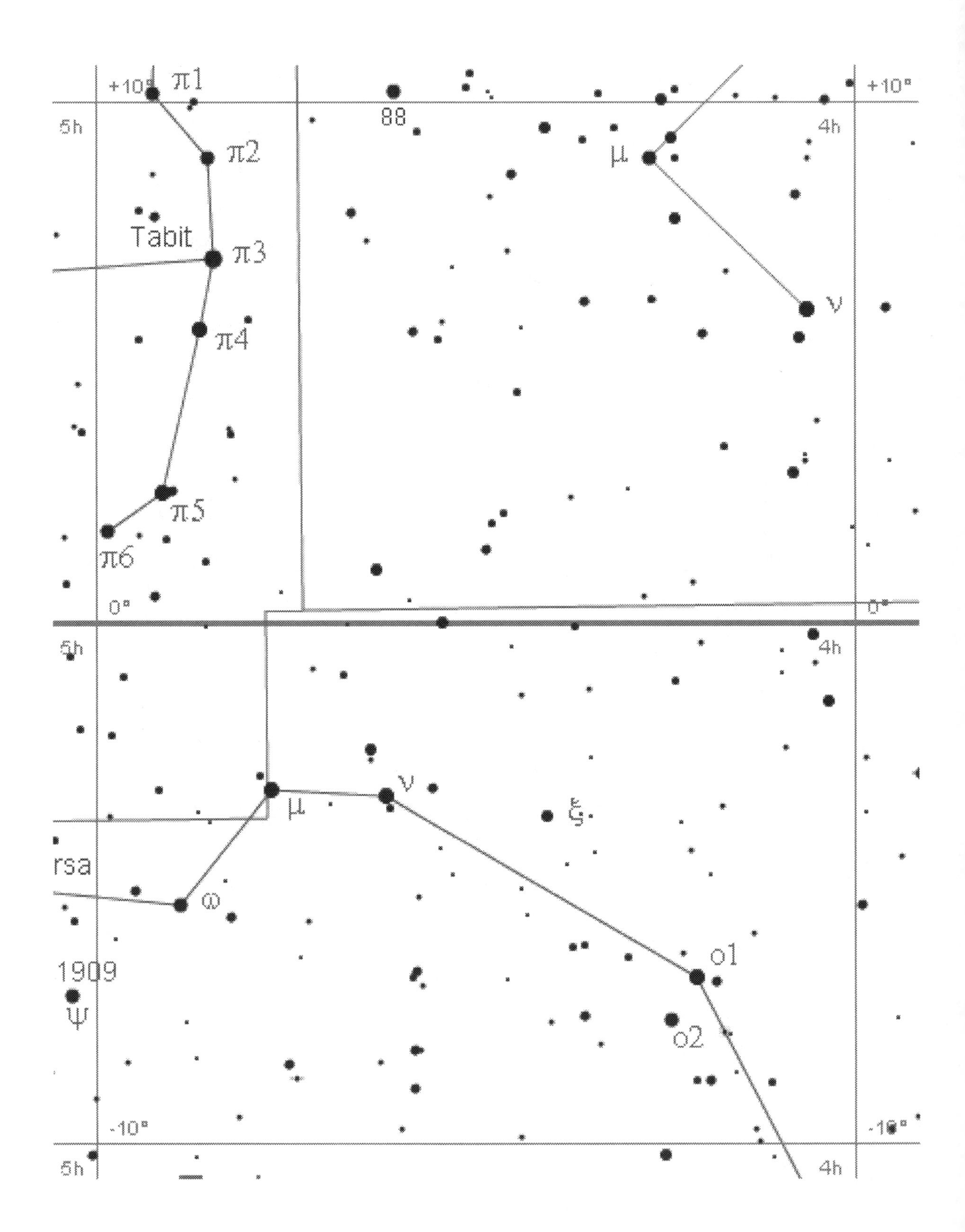

+10°
π1
88
+10°
5h
4h
π2
μ
Tabit
π3
ν
π4
π5
π6
0°
0°
5h
4h
ν
μ
ξ
rsa
ω
o1
1909
ψ
o2
-10°
-10°
5h
4h

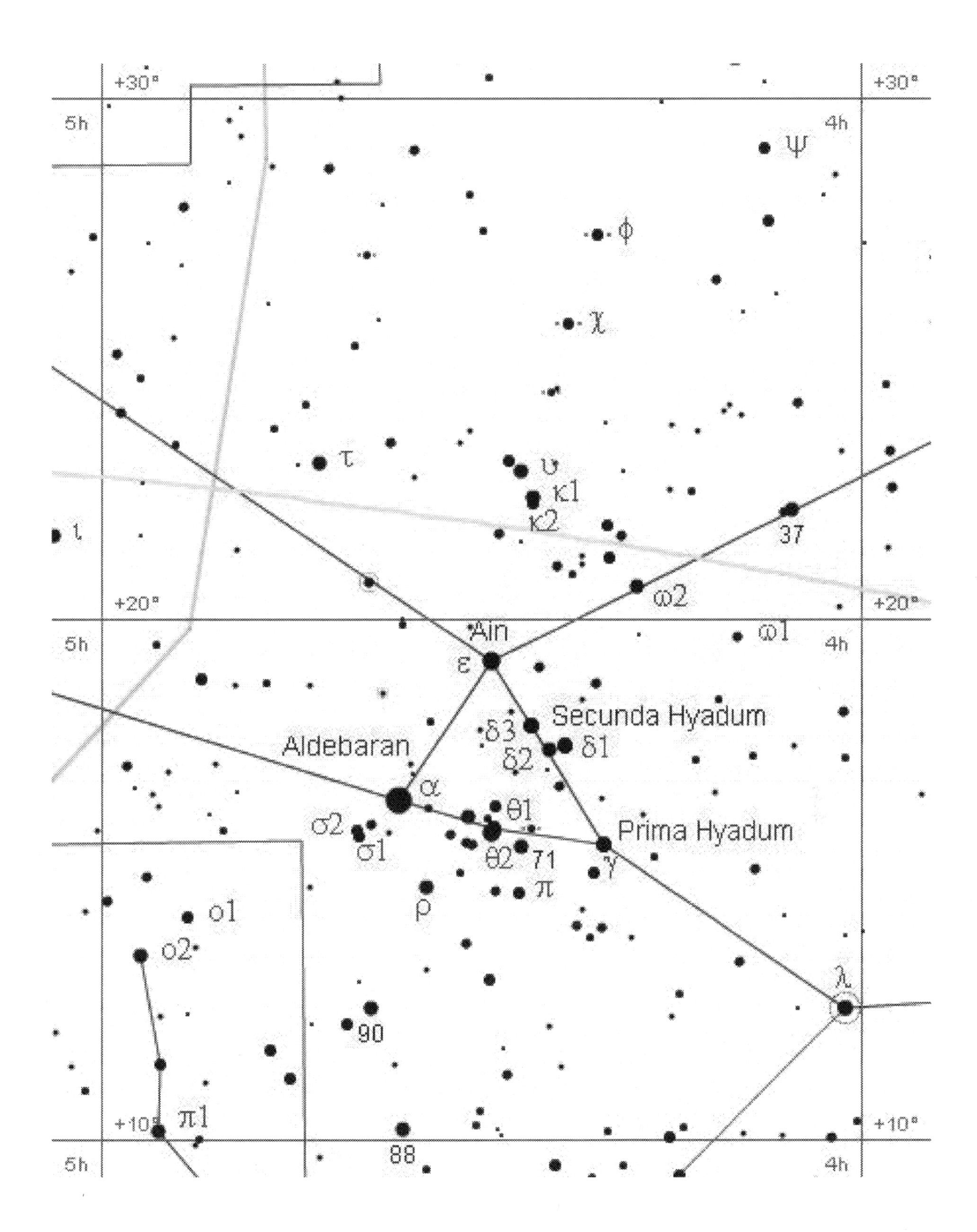

+30°
5h
4h
ψ
φ
χ
τ
υ
κ1
κ2
ι
37
ω2
+20°
Ain
ω1
ε
Secunda Hyadum
δ3
δ1
Aldebaran
δ2
α
θ1
σ2
Prima Hyadum
σ1
θ2
71
γ
π
ρ
ο1
ο2
λ
90
+10°
π1
88

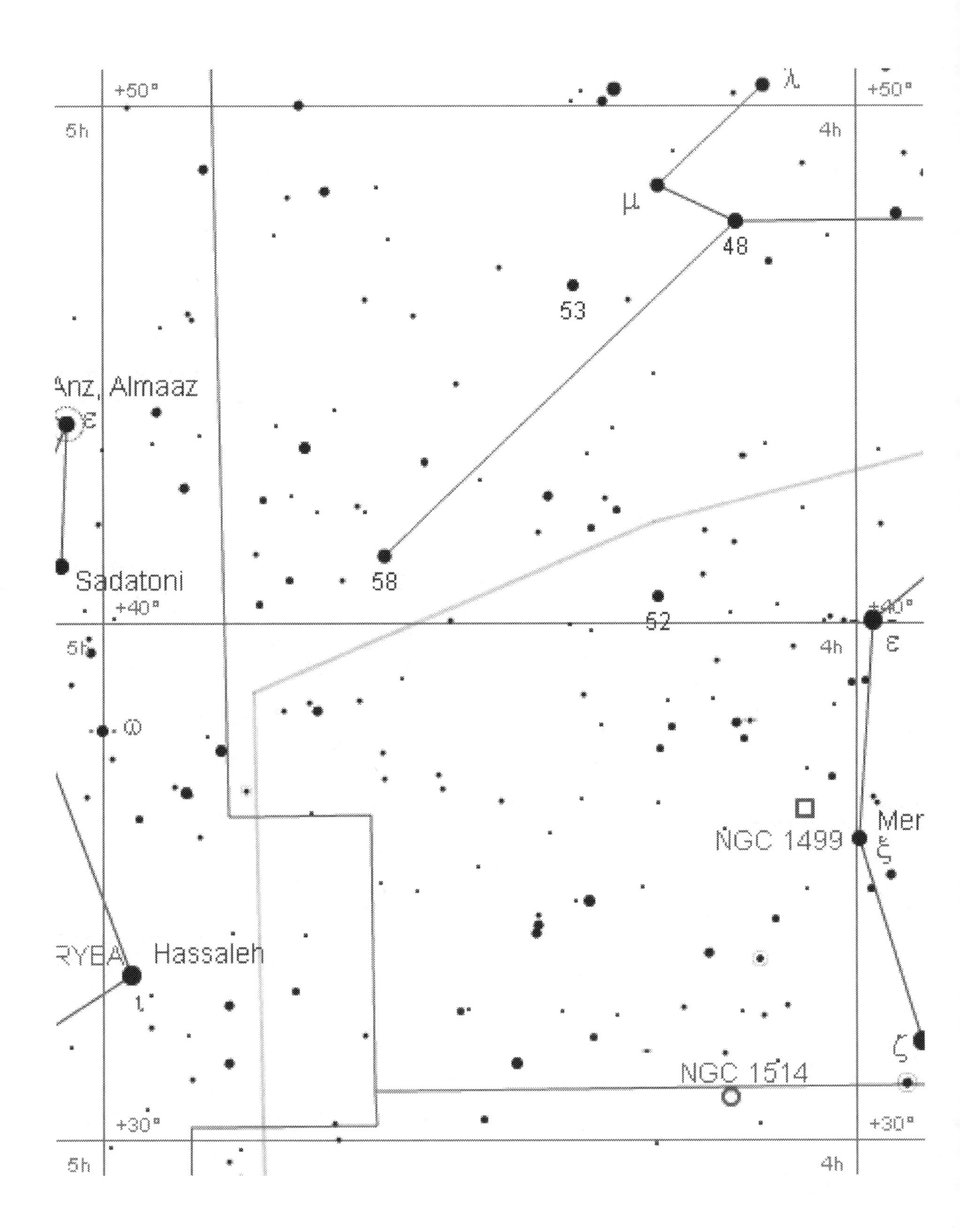
+50°
5h
4h
λ
μ
48
53
Anz, Almaaz
ε
Sadatoni
58
52
+40°
ω
Mer
NGC 1499
ξ
RYEA
Hassaleh
ι
ζ
NGC 1514
+30°

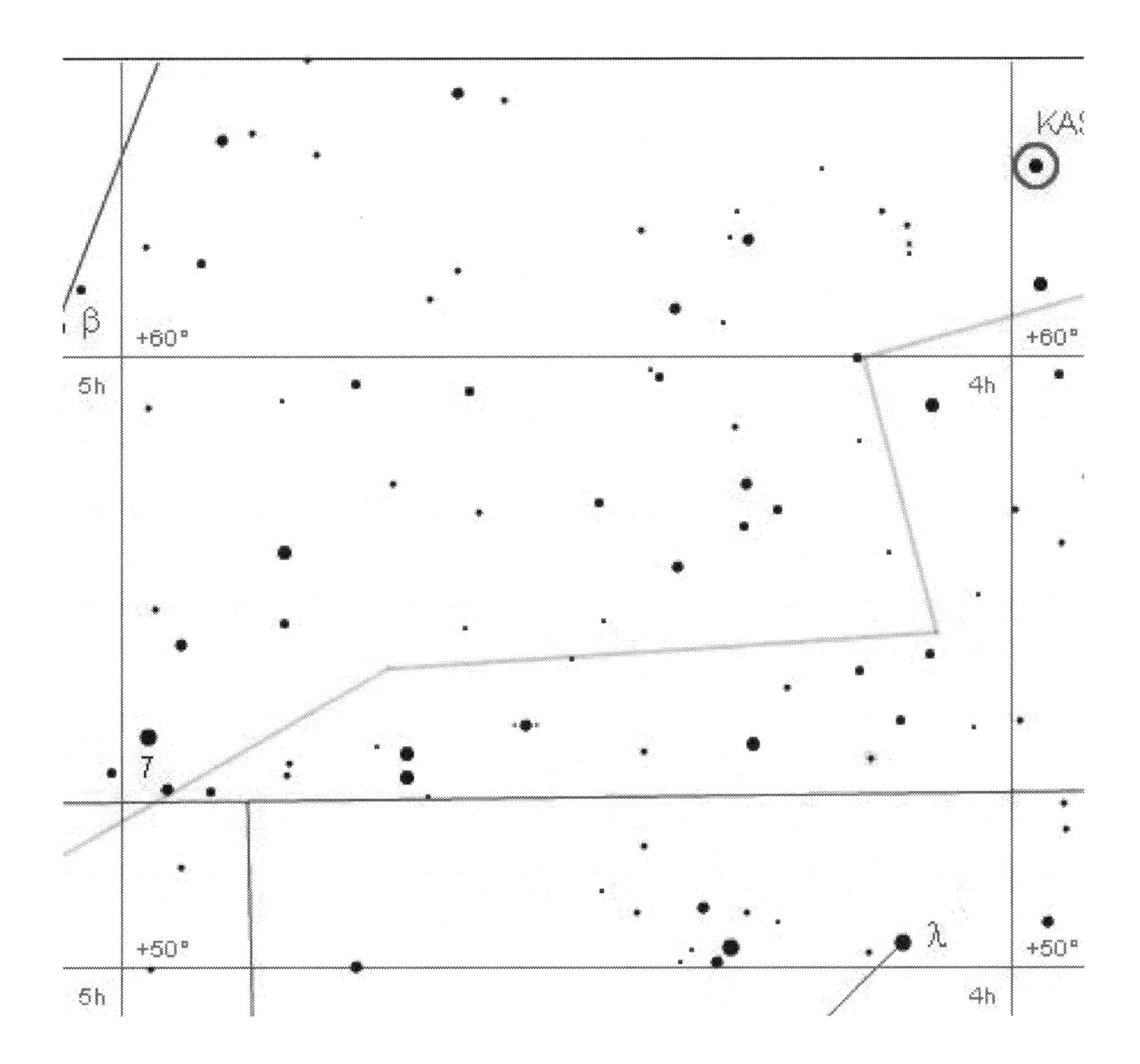
KAS
β
+60°
+60°
5h
4h
7
+50°
λ
+50°
5h
4h

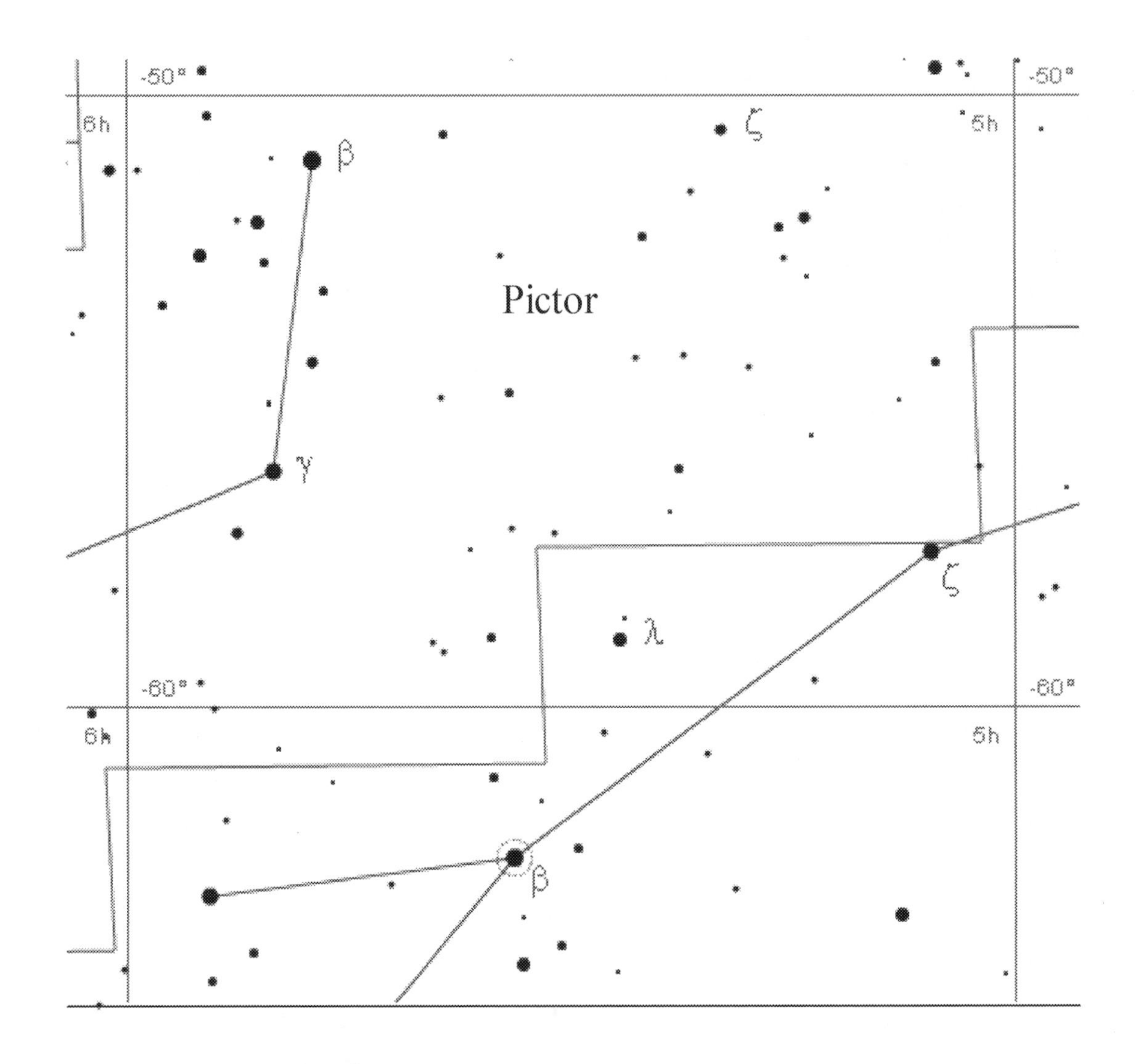
-50°
6h
5h
β
ζ
Pictor
γ
ζ
λ
-60°
6h
5h
β

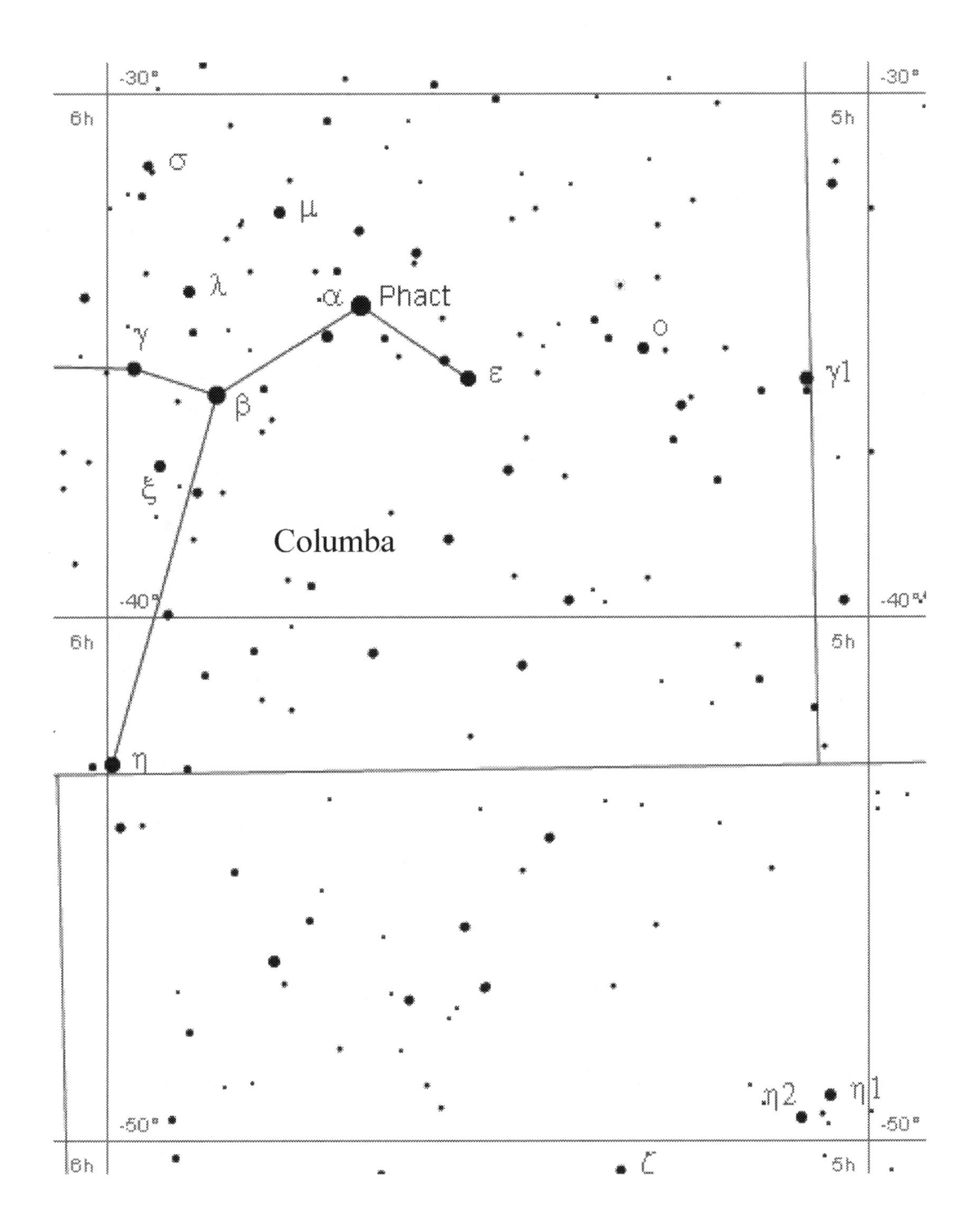
-30°
6h
5h
σ
μ
λ
α
Phact
γ
o
ε
γ1
β
ξ
Columba
-40°
η
η2
η1
-50°
ζ

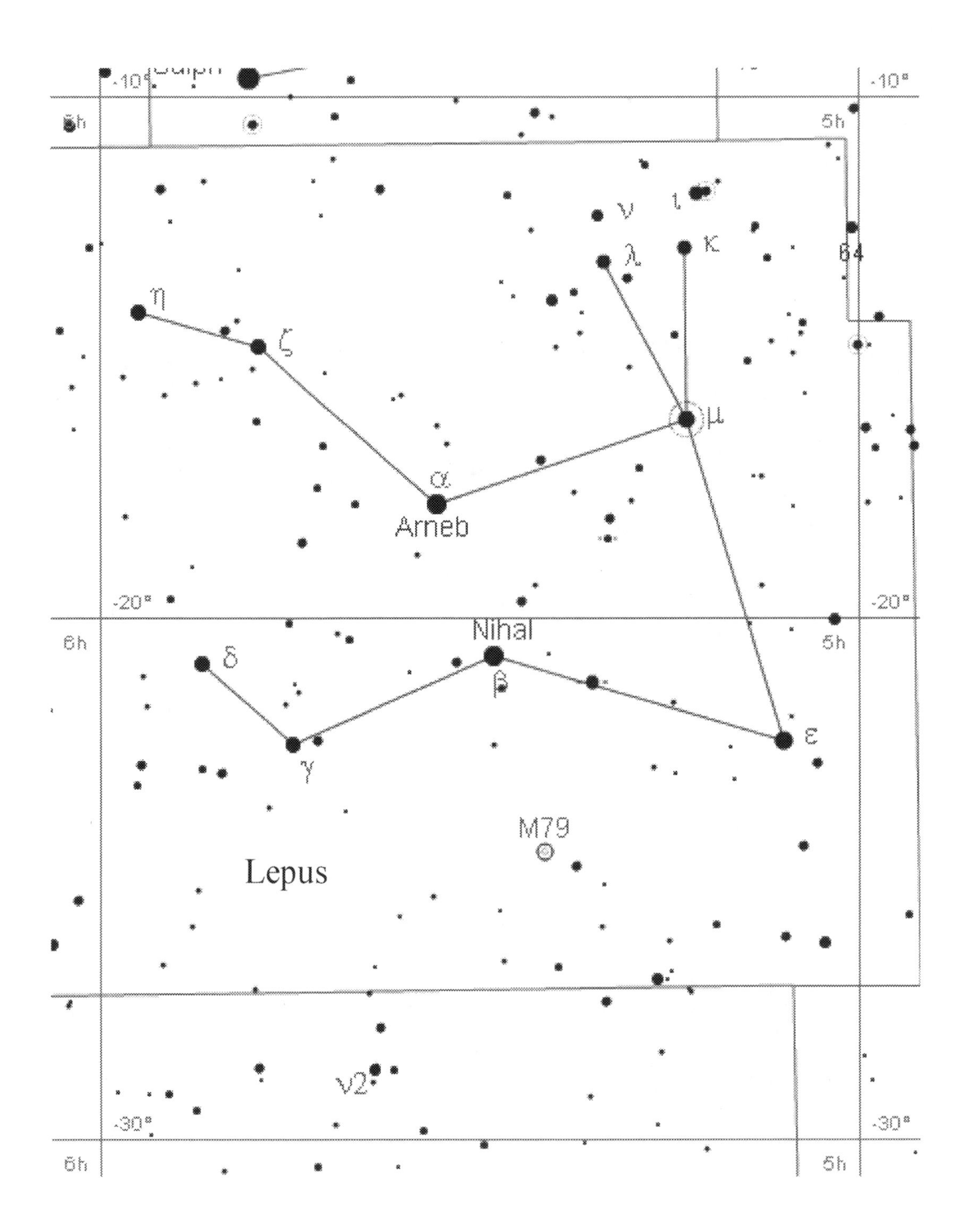
-10°
5h
ν
ι
λ
κ
64
η
ζ
μ
α
Arneb
-20°
6h
Nihal
δ
β
ε
γ
M79
Lepus
ν2
-30°

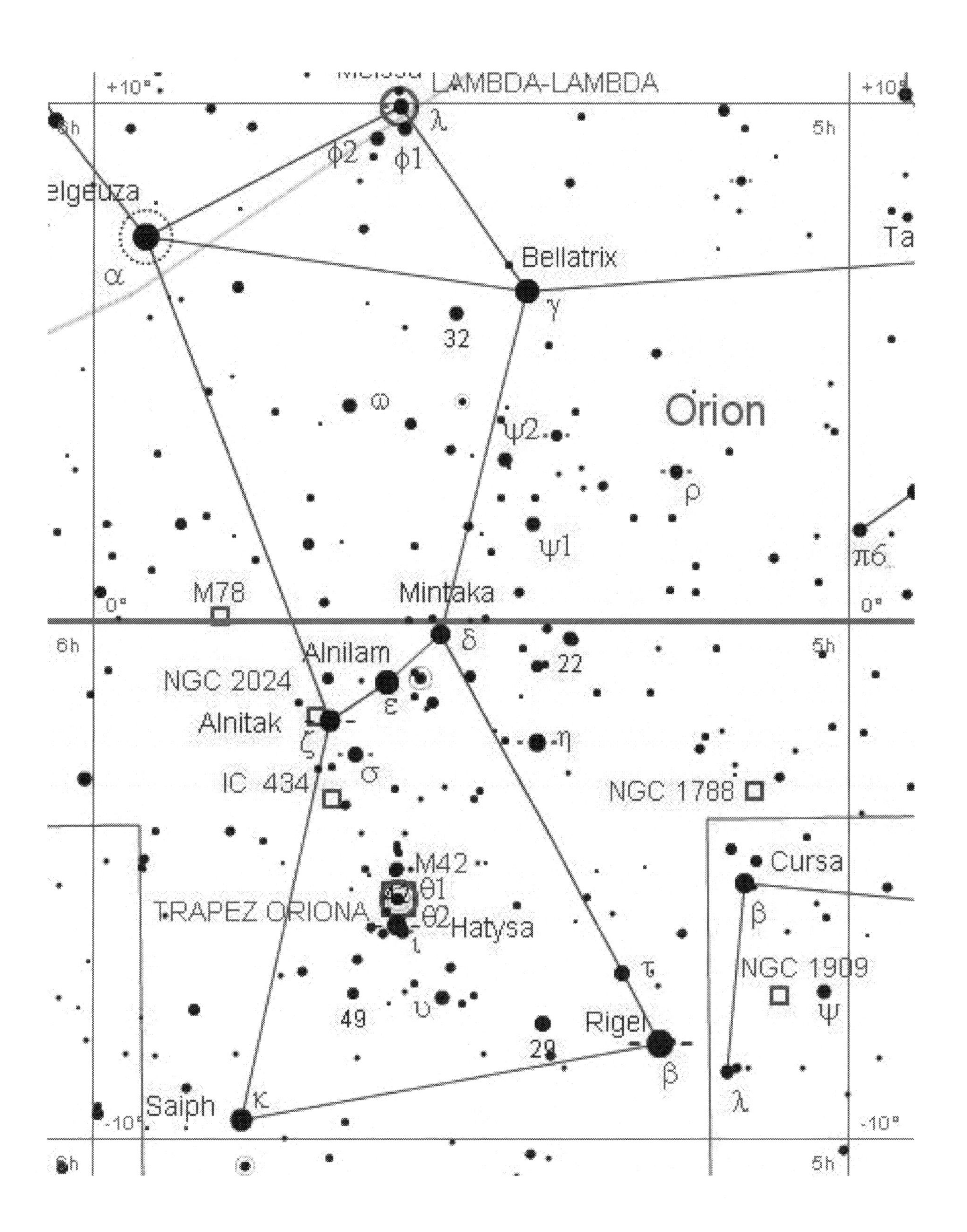
LAMBDA-LAMBDA
Orion
Bellatrix
Mintaka
Alnilam
NGC 2024
Alnitak
IC 434
M78
NGC 1788
M42
TRAPEZ ORIONA
Hatysa
Cursa
NGC 1909
Rigel
Saiph
32
22
49
29
+10°
0°
-10°
5h
6h

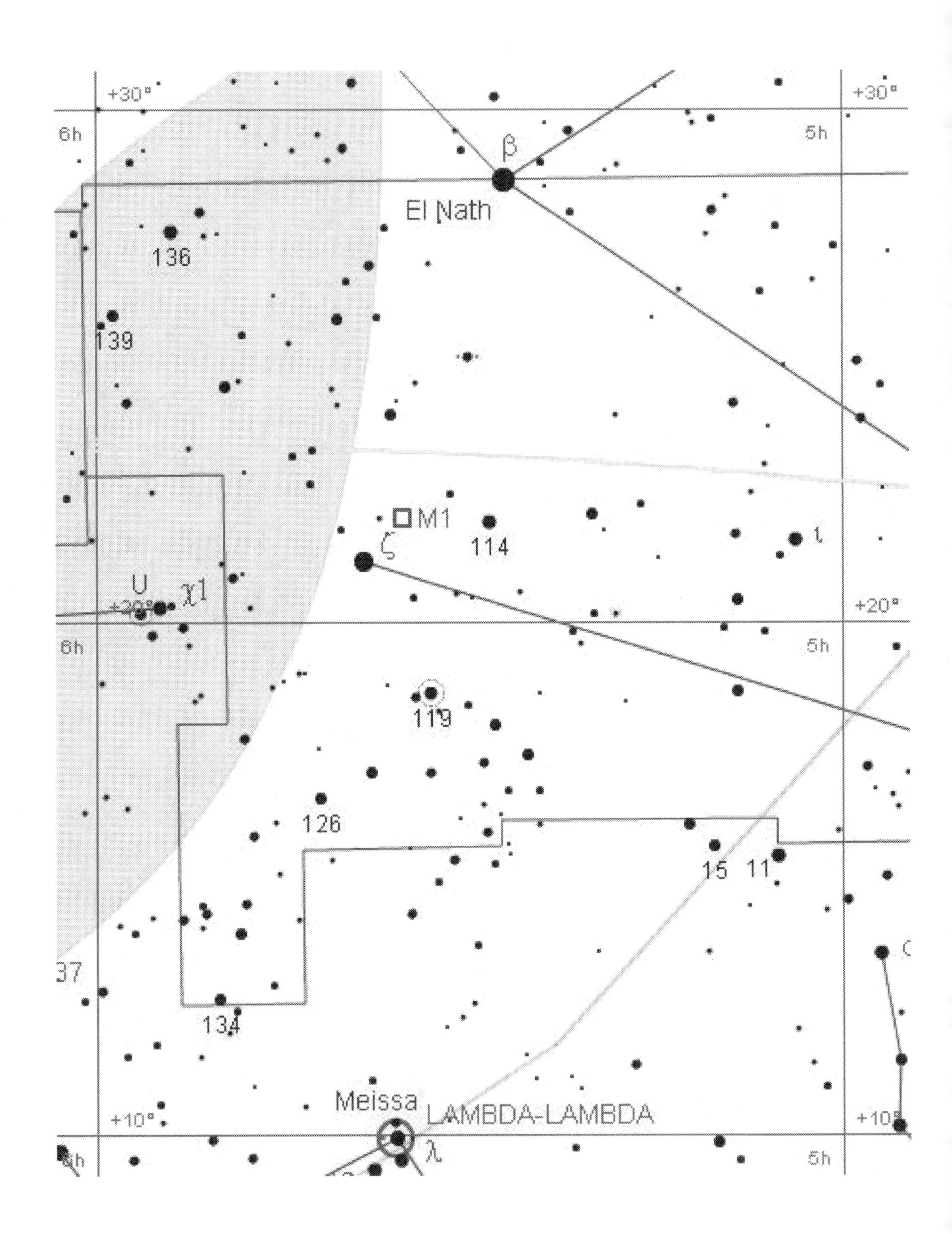
+30°
6h
5h
β
El Nath
136
139
M1
ζ
114
ι
U
χ1
+20°
119
126
15
11
37
134
Meissa
LAMBDA-LAMBDA
λ
+10°

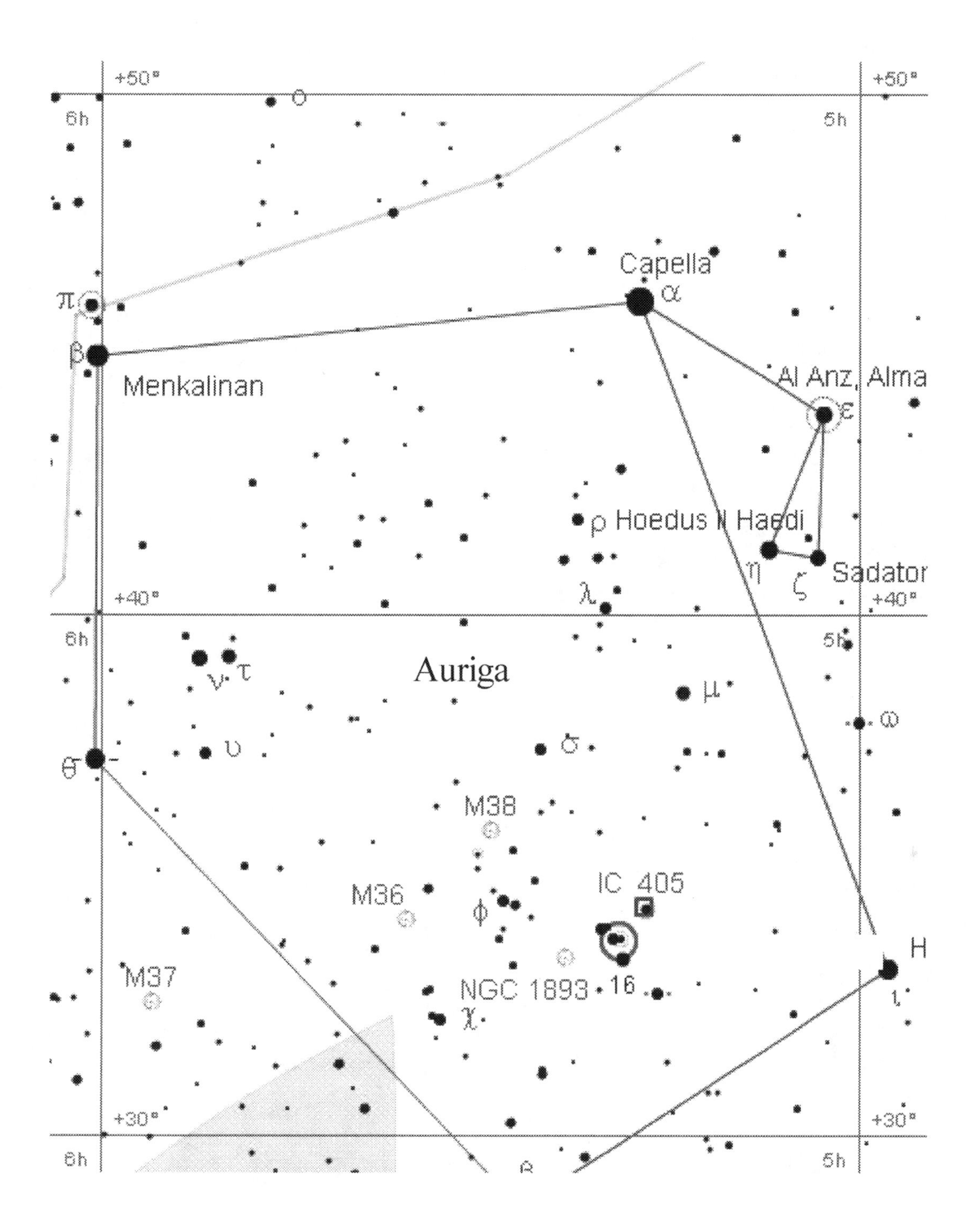
+50°
6h
5h
Capella
α
π
β
Menkalinan
Al Anz, Alma
ε
ρ Hoedus II Haedi
η
ζ
Sadator
λ
+40°
Auriga
ν
τ
μ
ω
υ
σ
θ
M38
M36
IC 405
φ
M37
NGC 1893
16
χ
ι
+30°

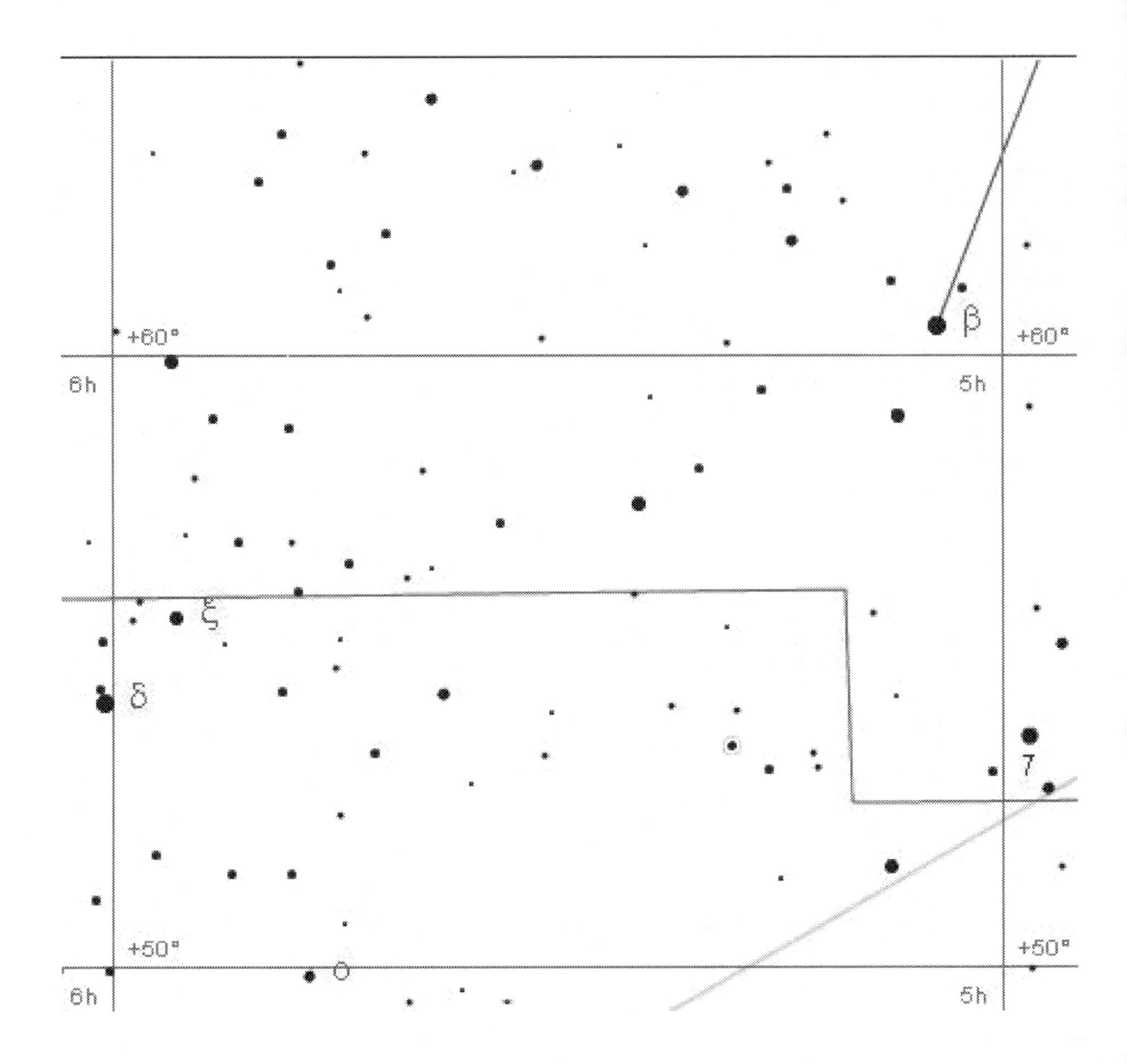
+60°
+60°
β
6h
5h
ξ
δ
7
+50°
+50°
o
6h
5h

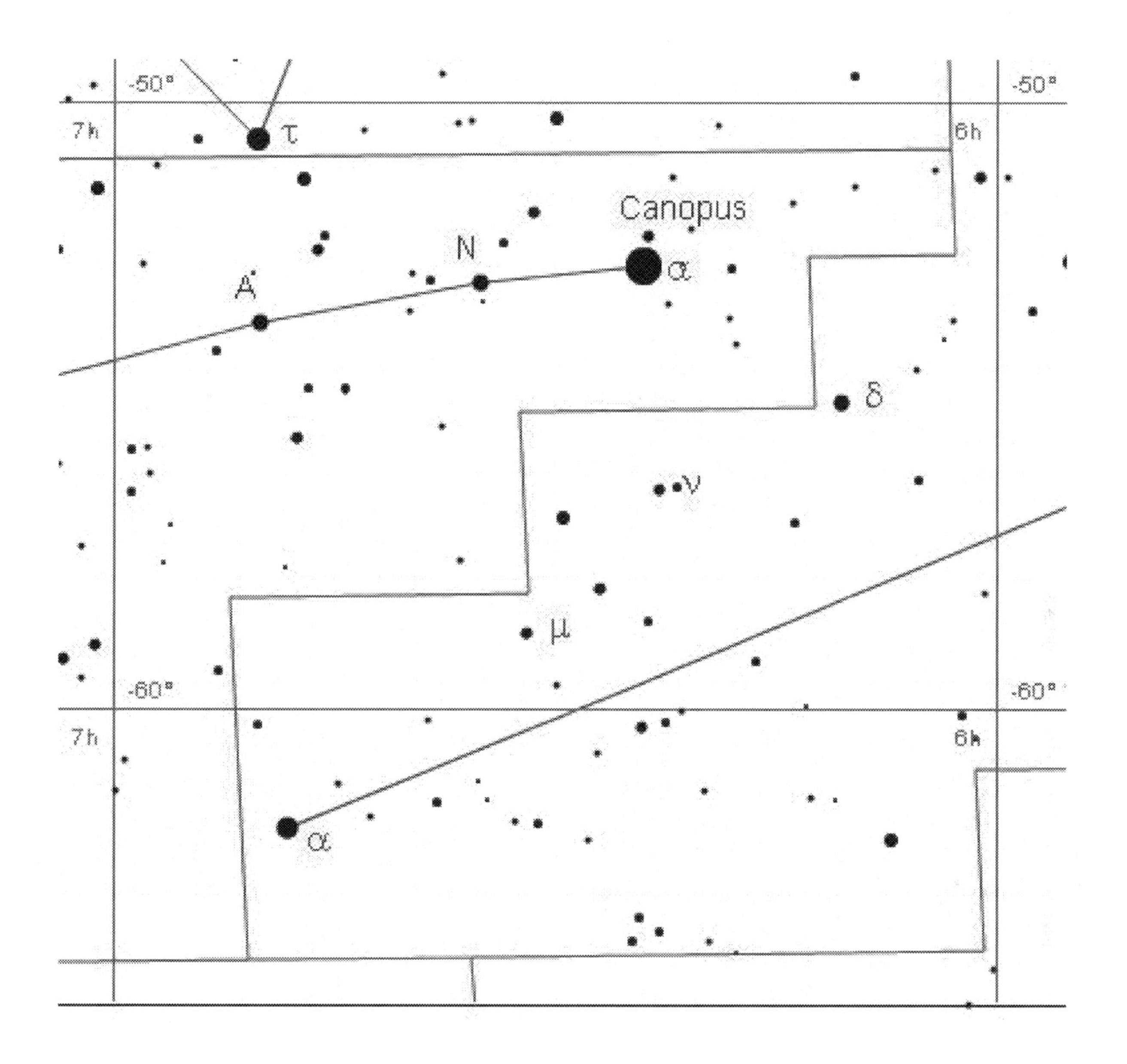
-50°
-50°
7h
6h
Canopus
N
A
α
δ
ν
μ
-60°
-60°
7h
6h
α

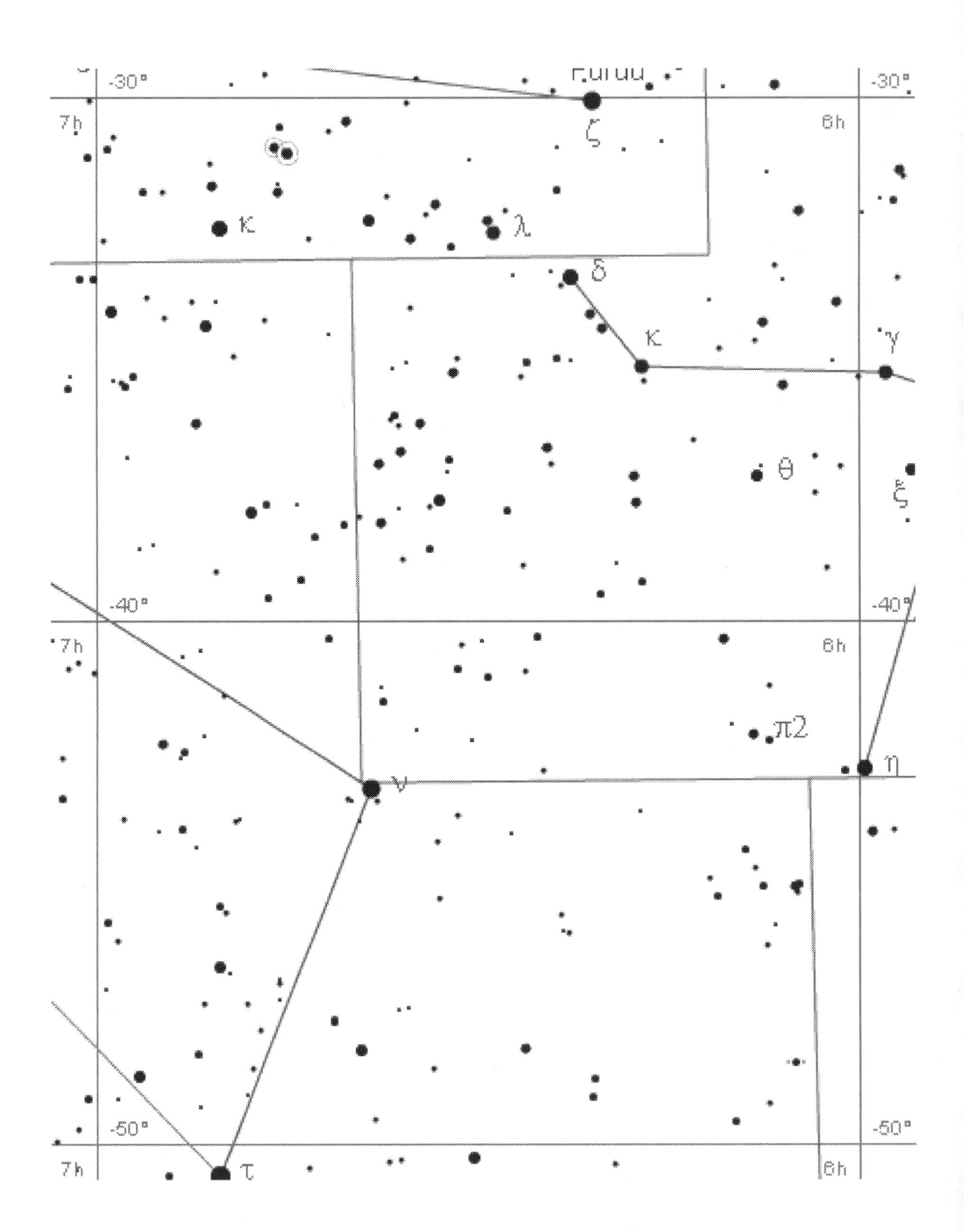
-30°
7h
6h
ζ
κ
λ
δ
κ
γ
θ
ξ
-40°
7h
6h
π2
η
ν
-50°
7h
τ
6h

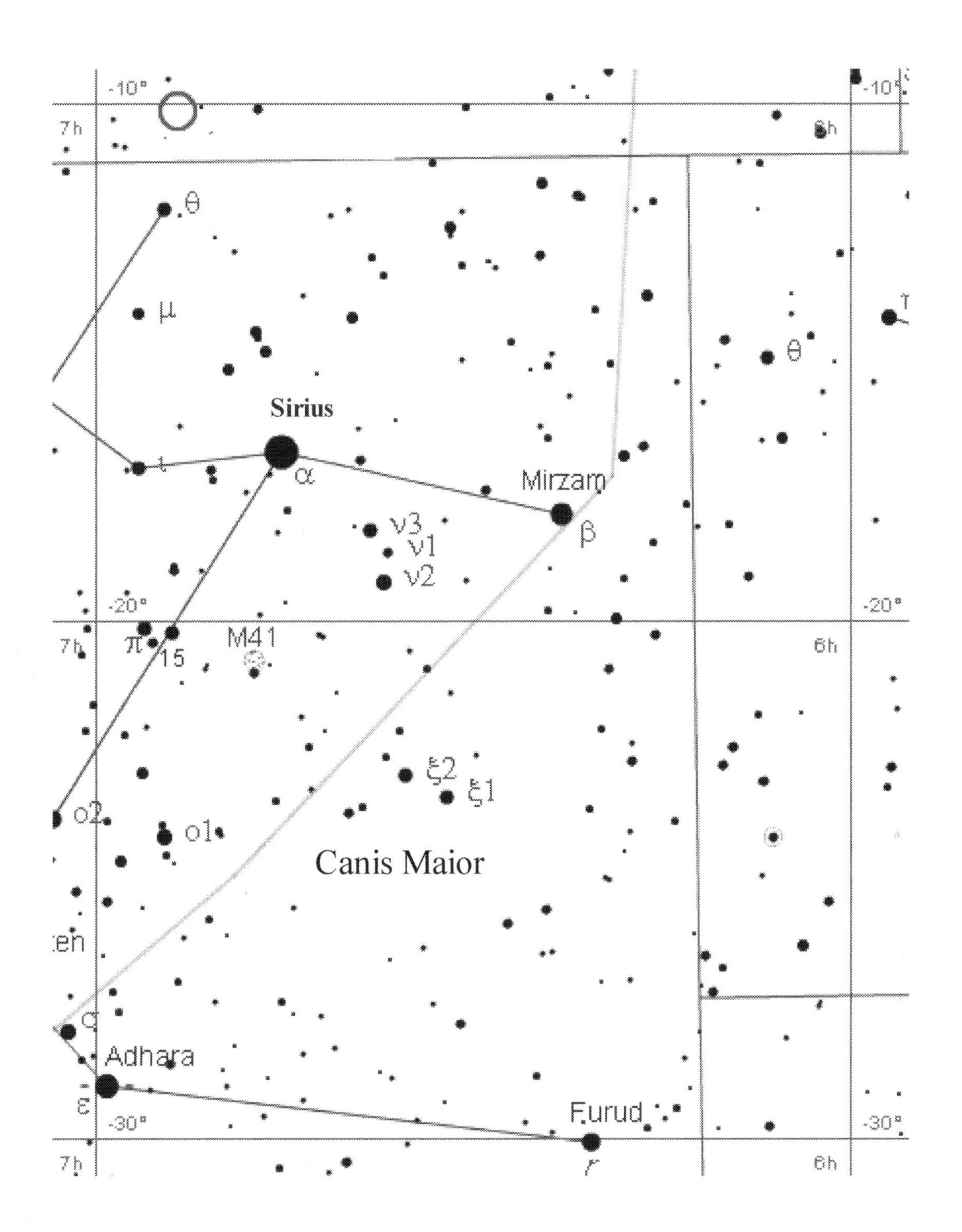
-10°
7h
6h
θ
μ
Sirius
α
Mirzam
β
ν3
ν1
ν2
θ
-20°
π
15
M41
ξ2
ξ1
o2
o1
Canis Maior
Adhara
ε
Furud
-30°

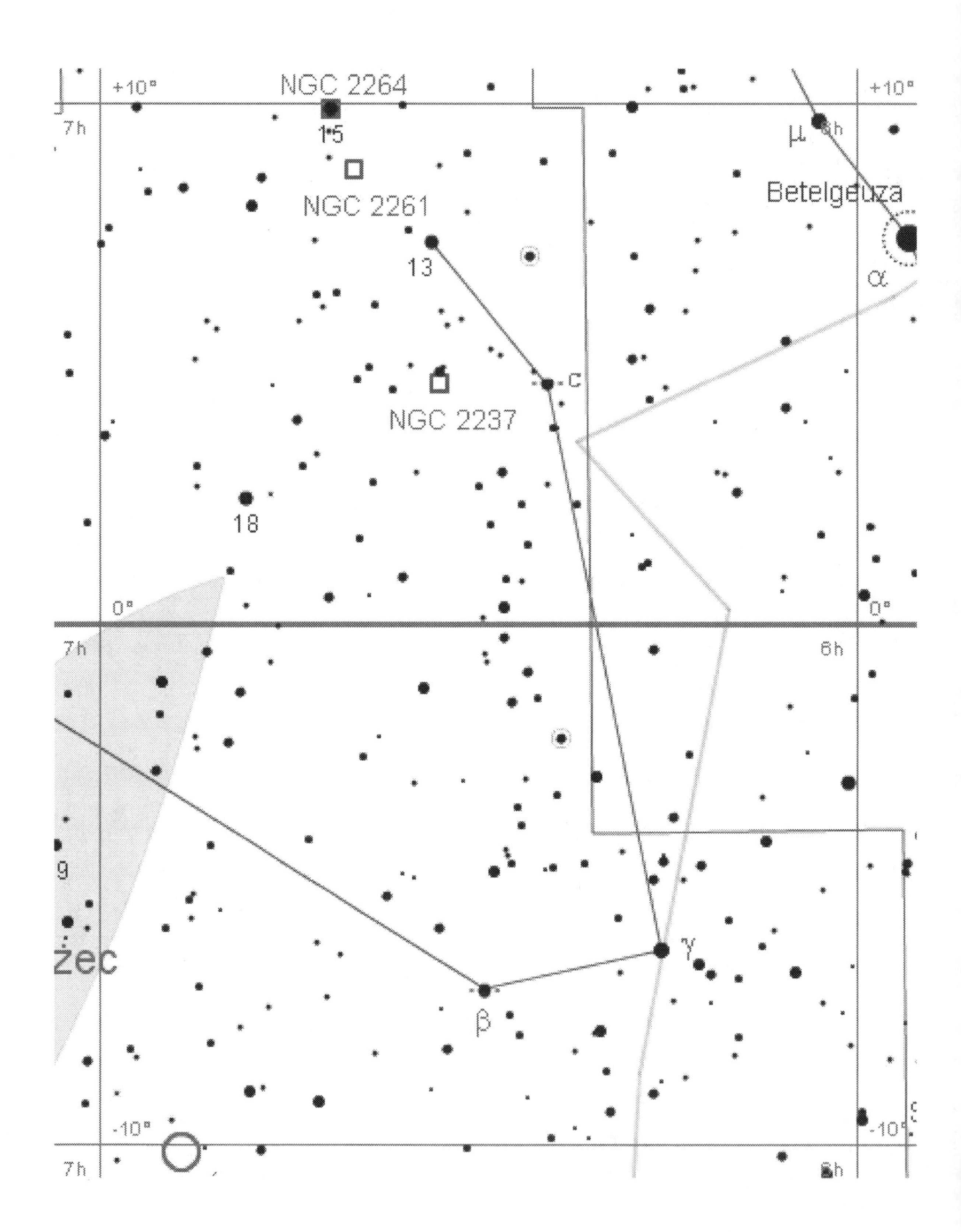
+10°
NGC 2264
15
NGC 2261
13
NGC 2237
c
18
0°
7h
6h
μ
Betelgeuza
α
γ
β
-10°
żec
9

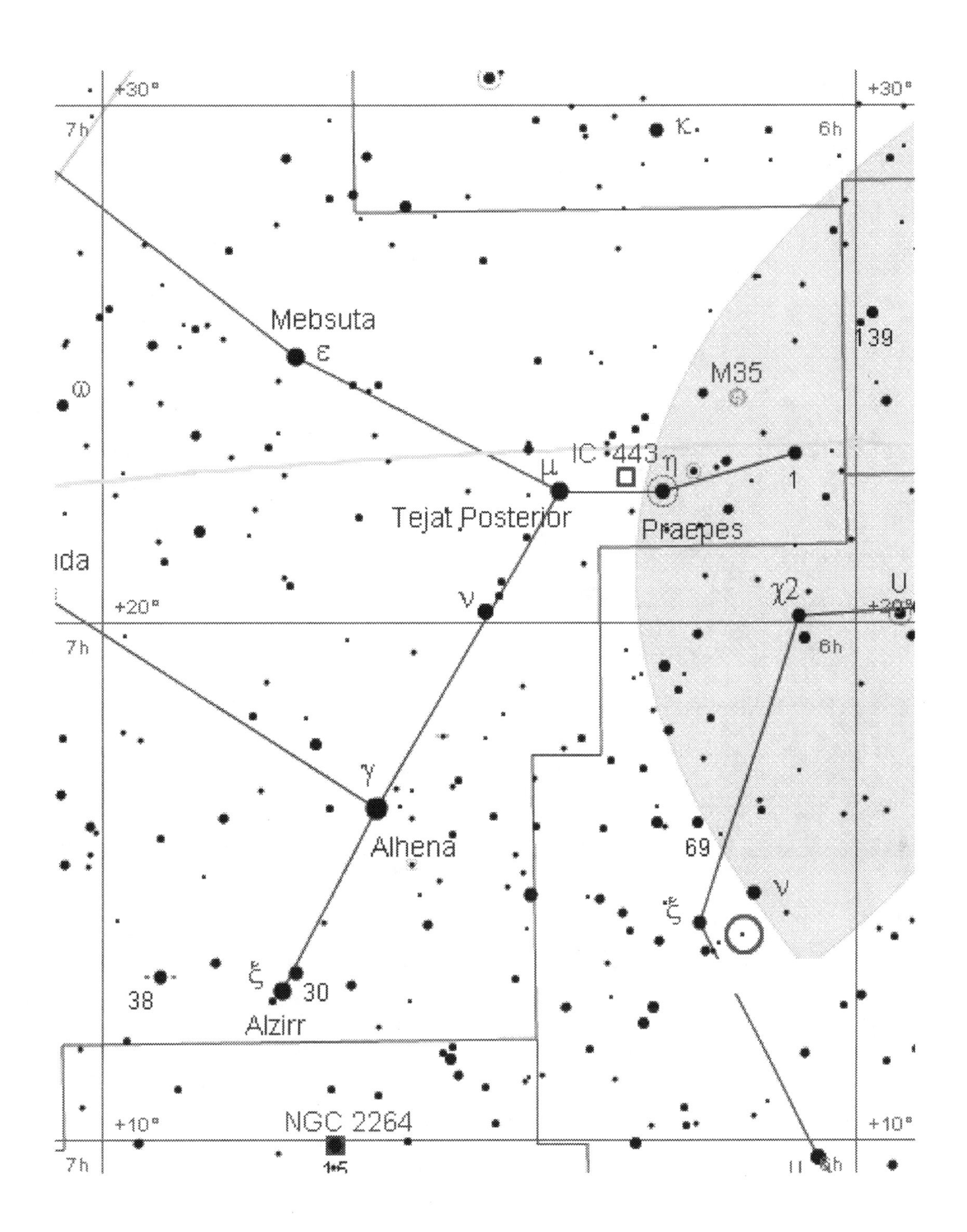
+30°
7h
κ
6h
Mebsuta
ε
ω
139
M35
IC 443
μ
η
1
Tejat Posterior
Praepes
ıda
ν
χ2
U
+20°
7h
6h
γ
Alhena
69
ν
ξ
ξ
30
38
Alzirr
NGC 2264
+10°
7h

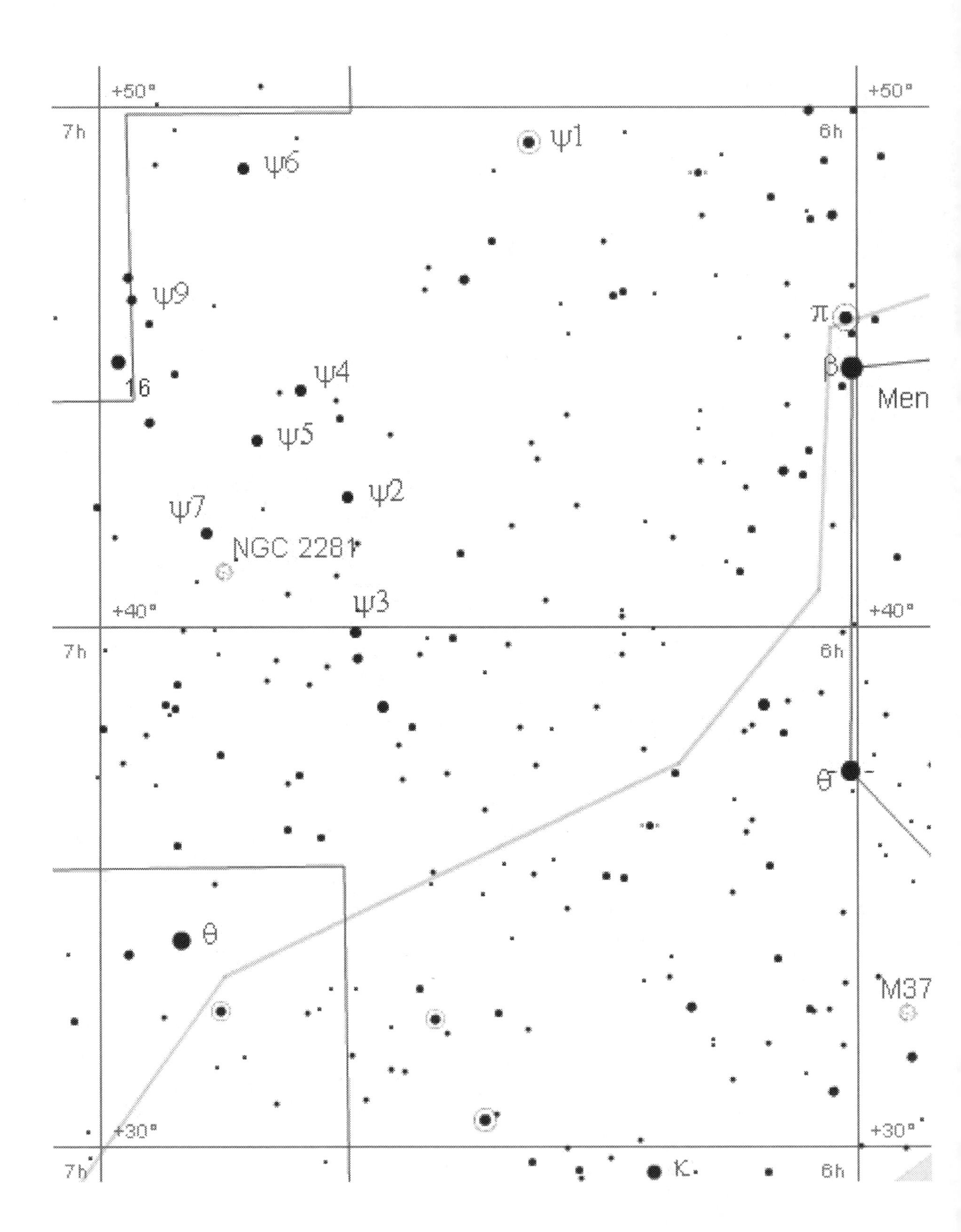
+50°
7h
6h
ψ1
ψ6
ψ9
π
β
Men
16
ψ4
ψ5
ψ2
ψ7
NGC 2281
ψ3
+40°
θ
M37
+30°
κ

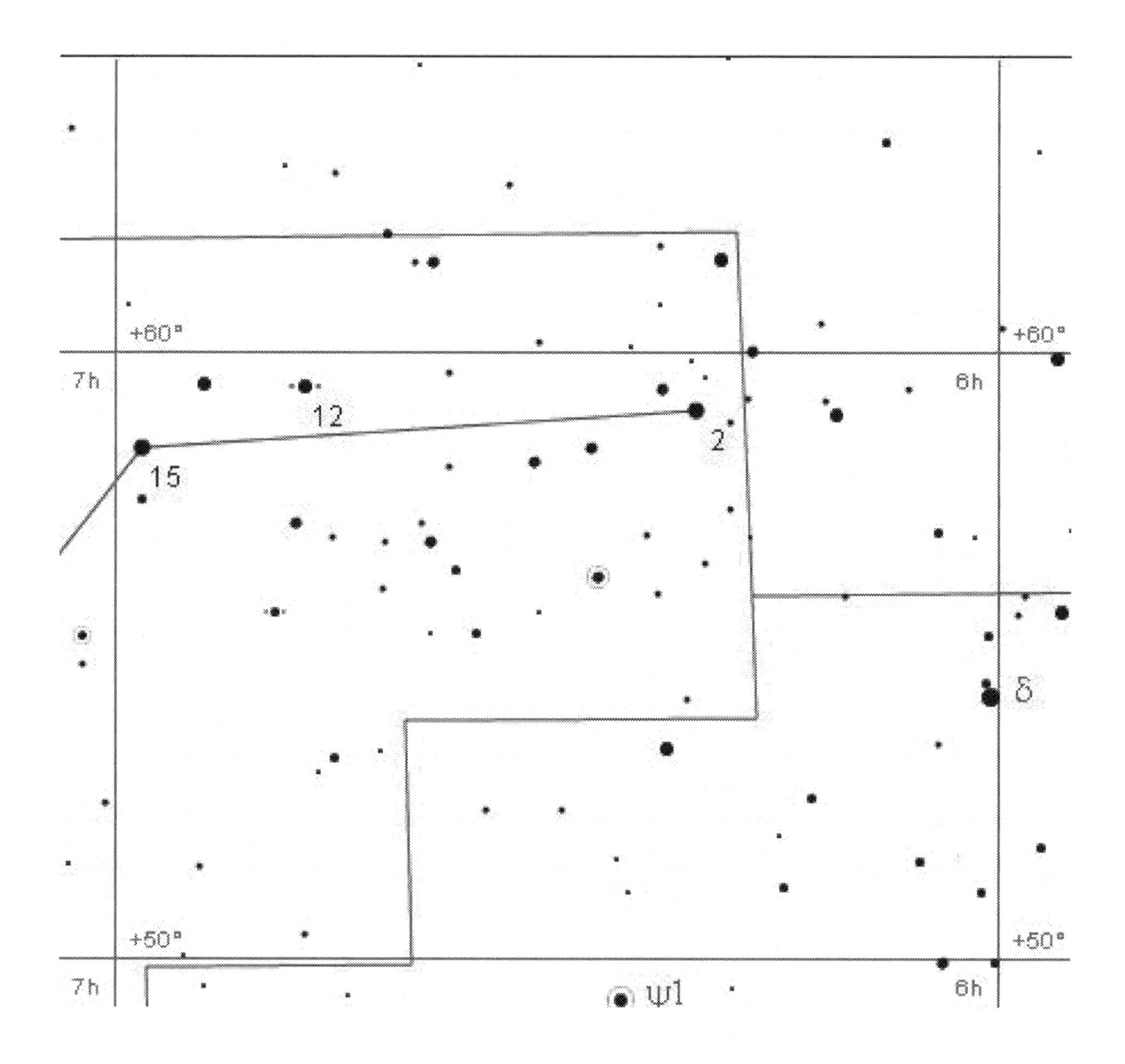
+60°
+60°
7h
12
6h
2
15
δ
+50°
+50°
7h
ψ1
6h

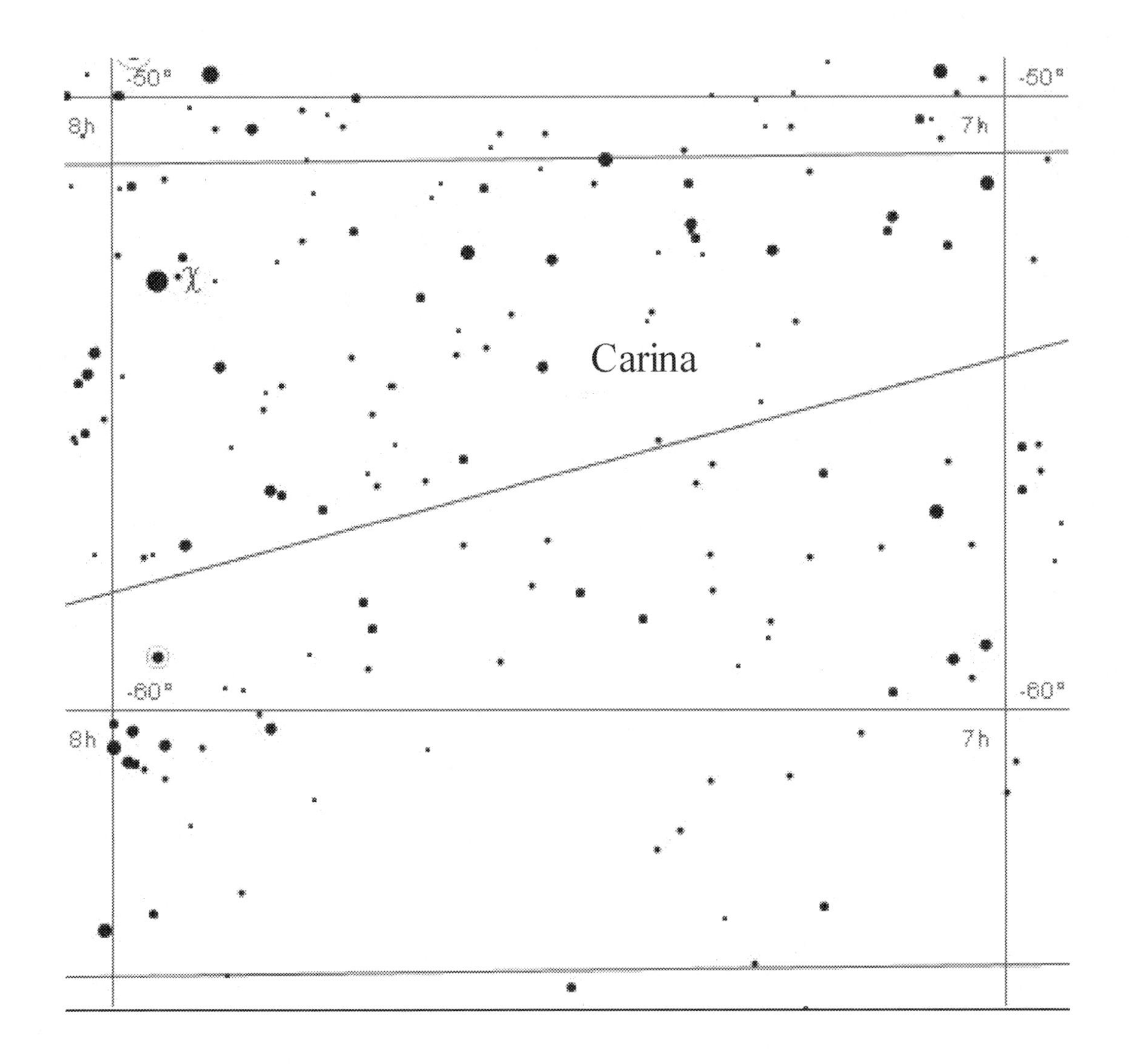
-50°
-50°
8h
7h
χ
Carina
-60°
-60°
8h
7h

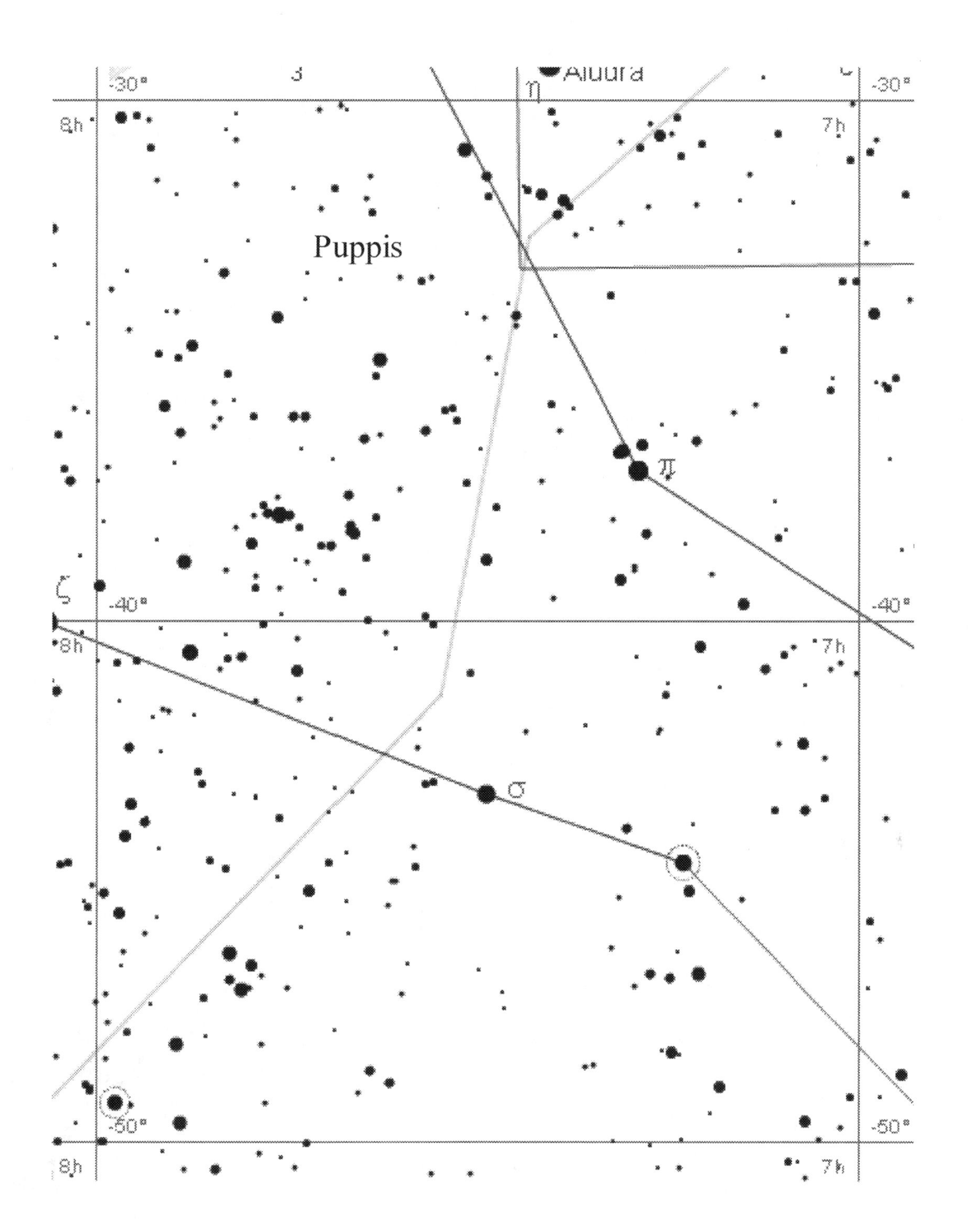

Aludra
η
-30°
8h
7h
Puppis
π
ζ
-40°
σ
-50°

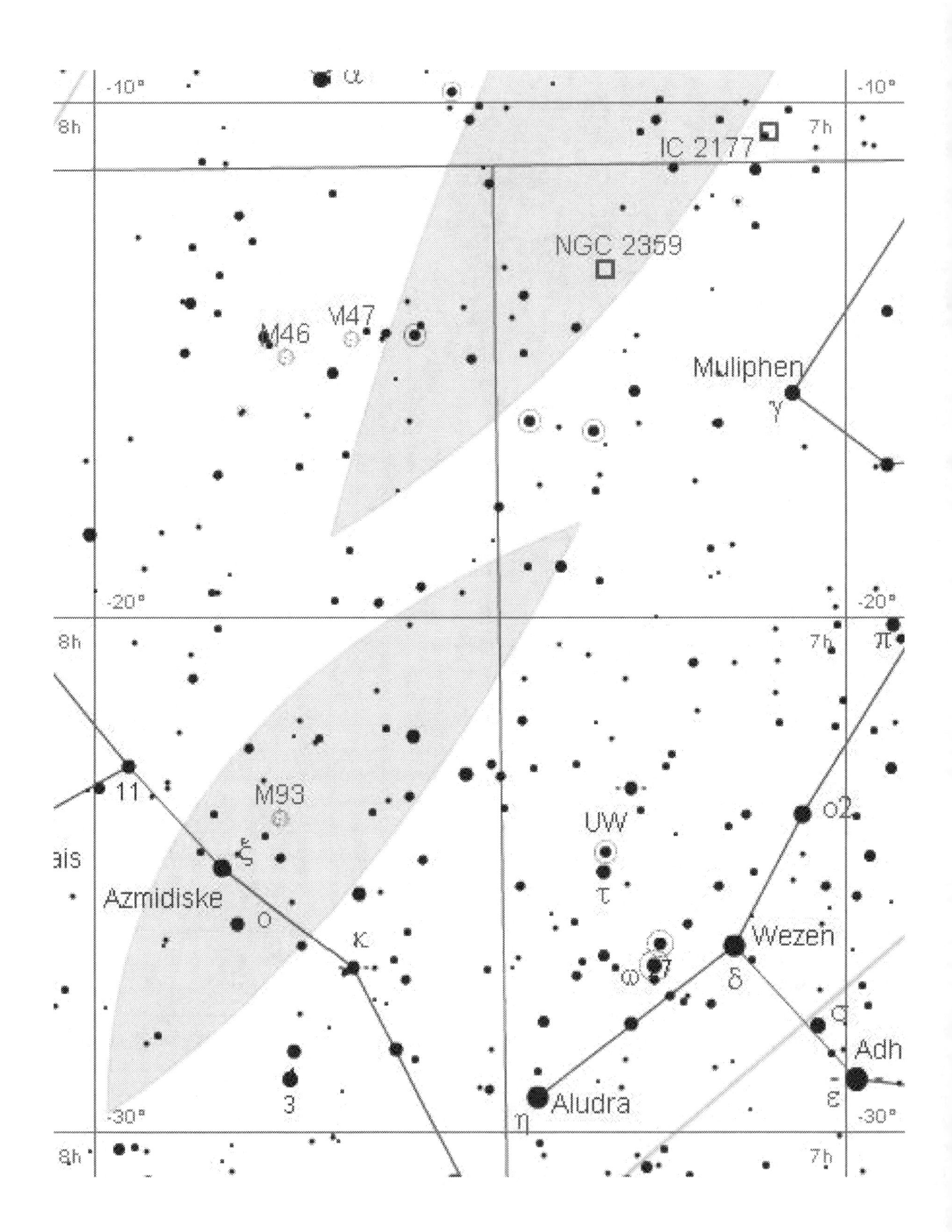

-10°
α
8h
7h
IC 2177
NGC 2359
M46
M47
Muliphen
γ
-20°
π
11
M93
UW
o2
ξ
ais
Azmidiske
τ
o
κ
Wezen
ω
27
δ
σ
Adh
3
Aludra
ε
-30°
η

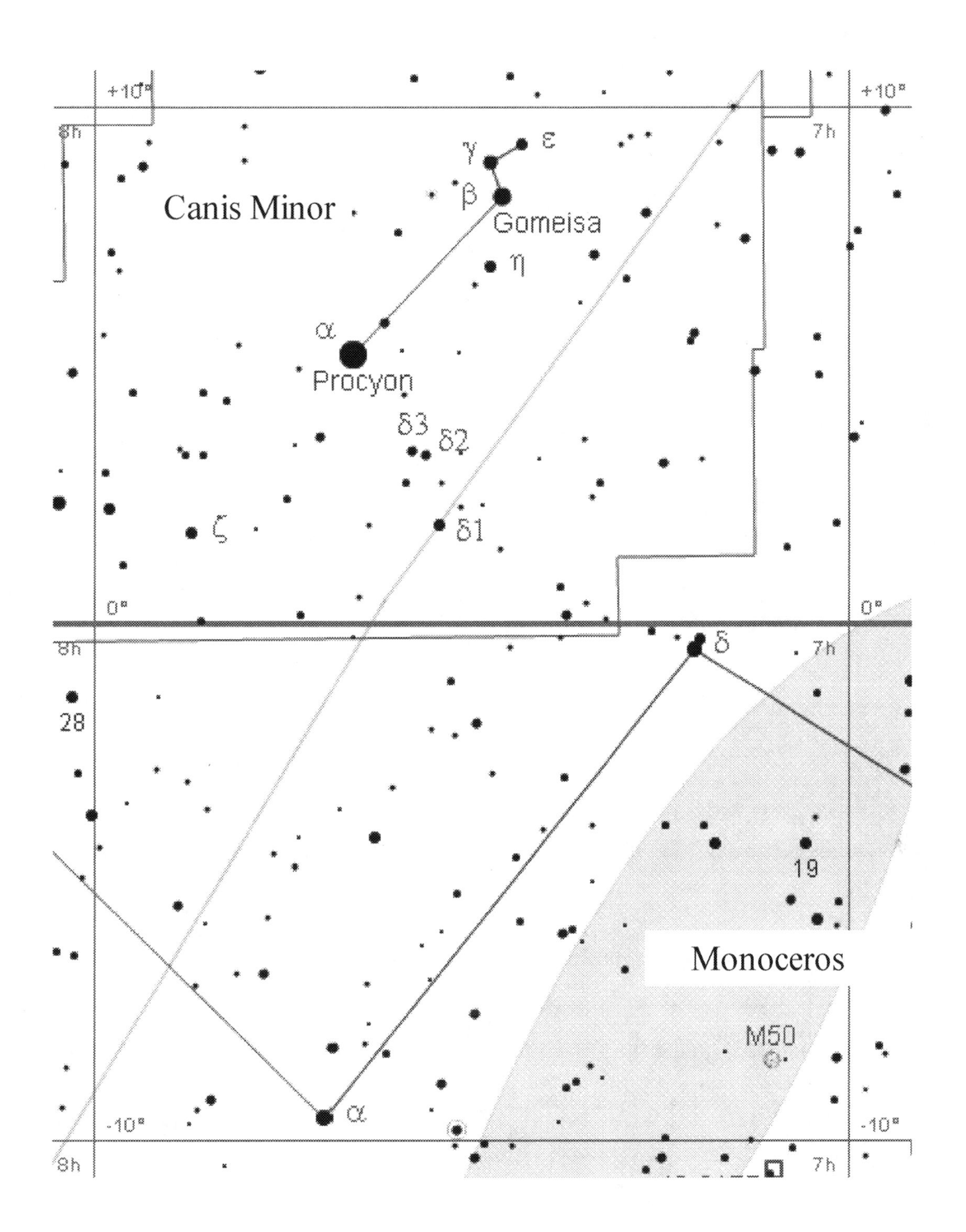
+10°
8h
7h
Canis Minor
γ
ε
β
Gomeisa
η
α
Procyon
δ3
δ2
δ1
ζ
0°
δ
28
19
Monoceros
M50
α
-10°

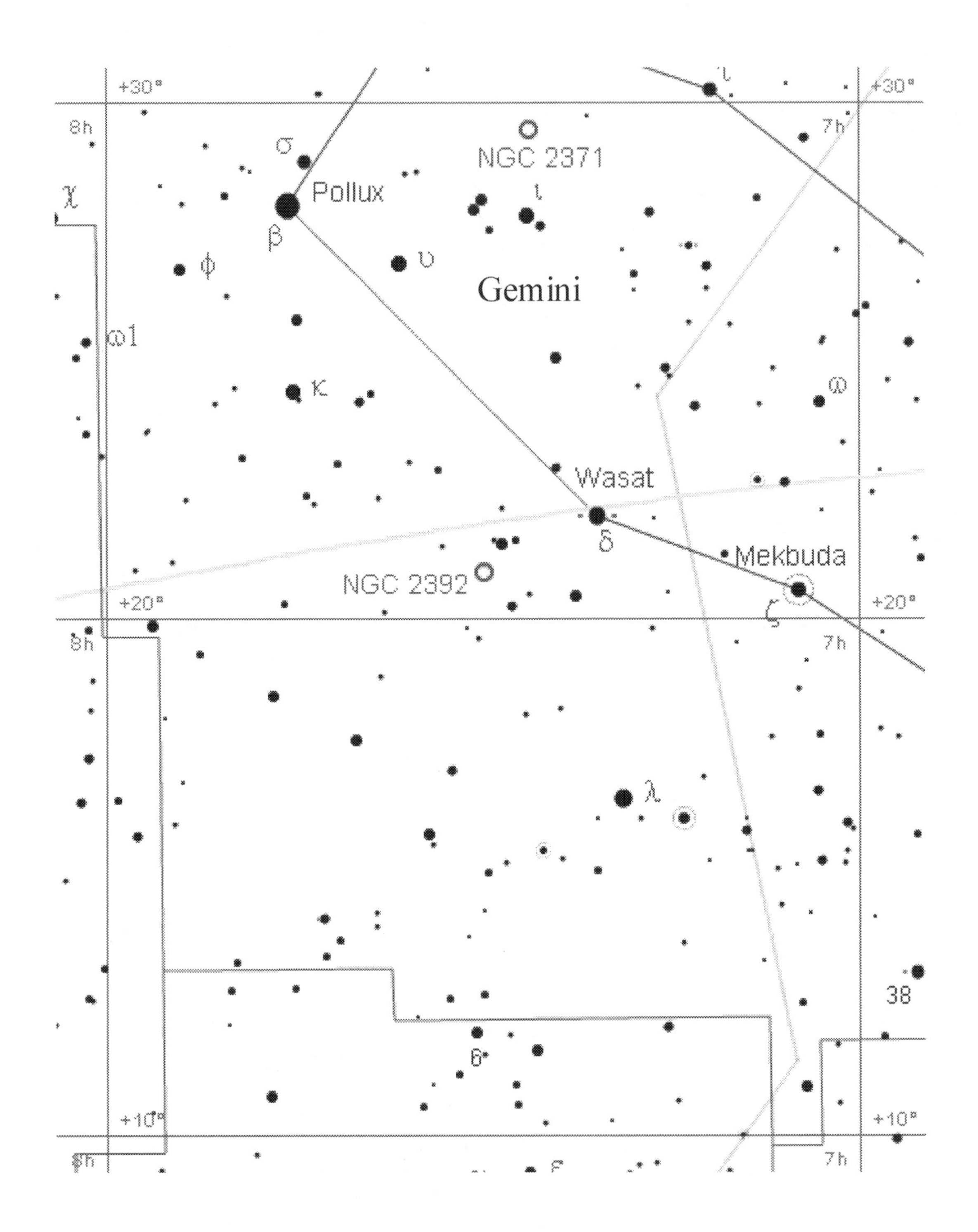
+30°
8h
7h
NGC 2371
Pollux
Gemini
Wasat
Mekbuda
NGC 2392
+20°
38
+10°

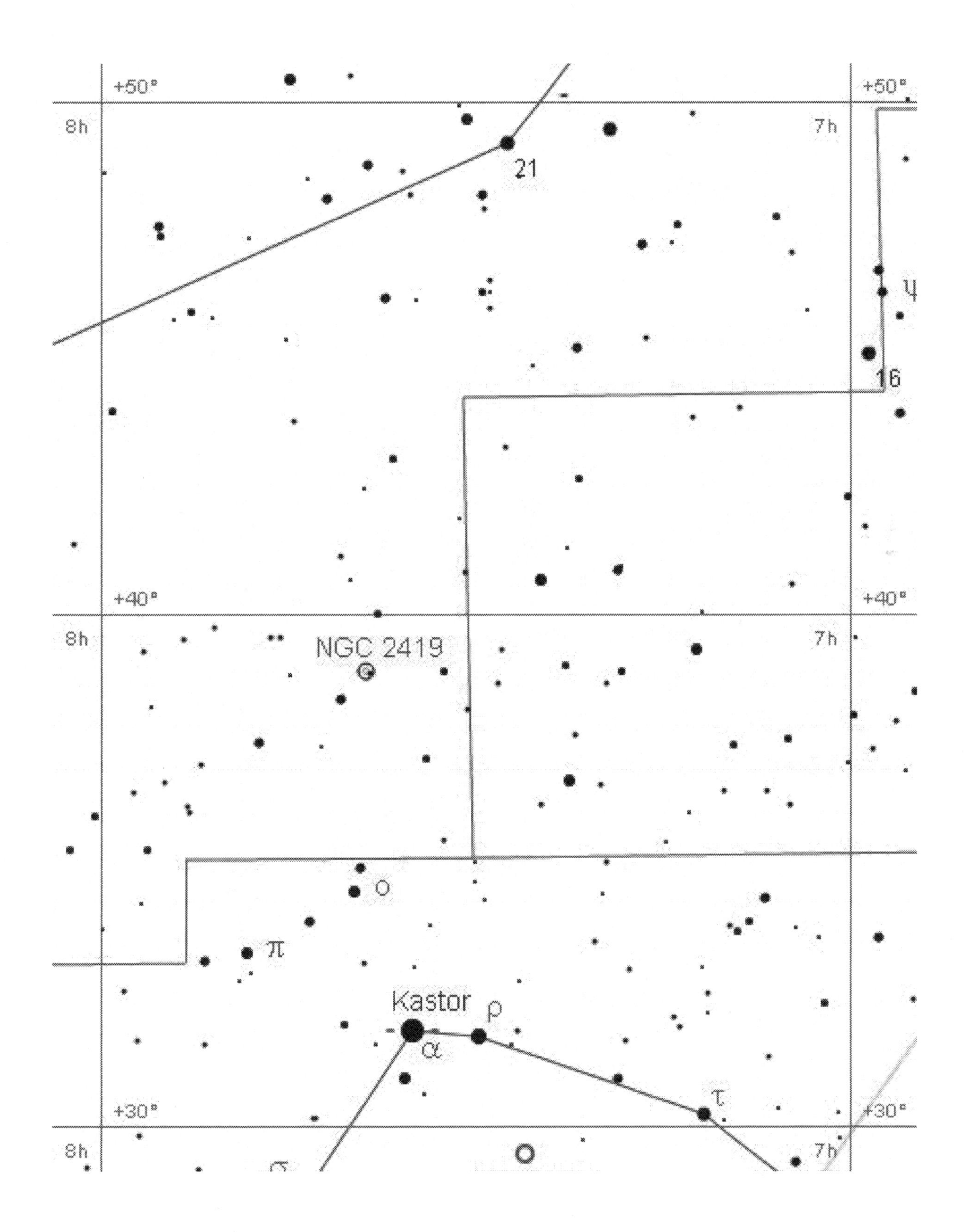
+50°
8h
21
7h
+50°
ψ
16
+40°
8h
NGC 2419
7h
+40°
o
π
Kastor
ρ
α
τ
+30°
8h
σ
7h
+30°

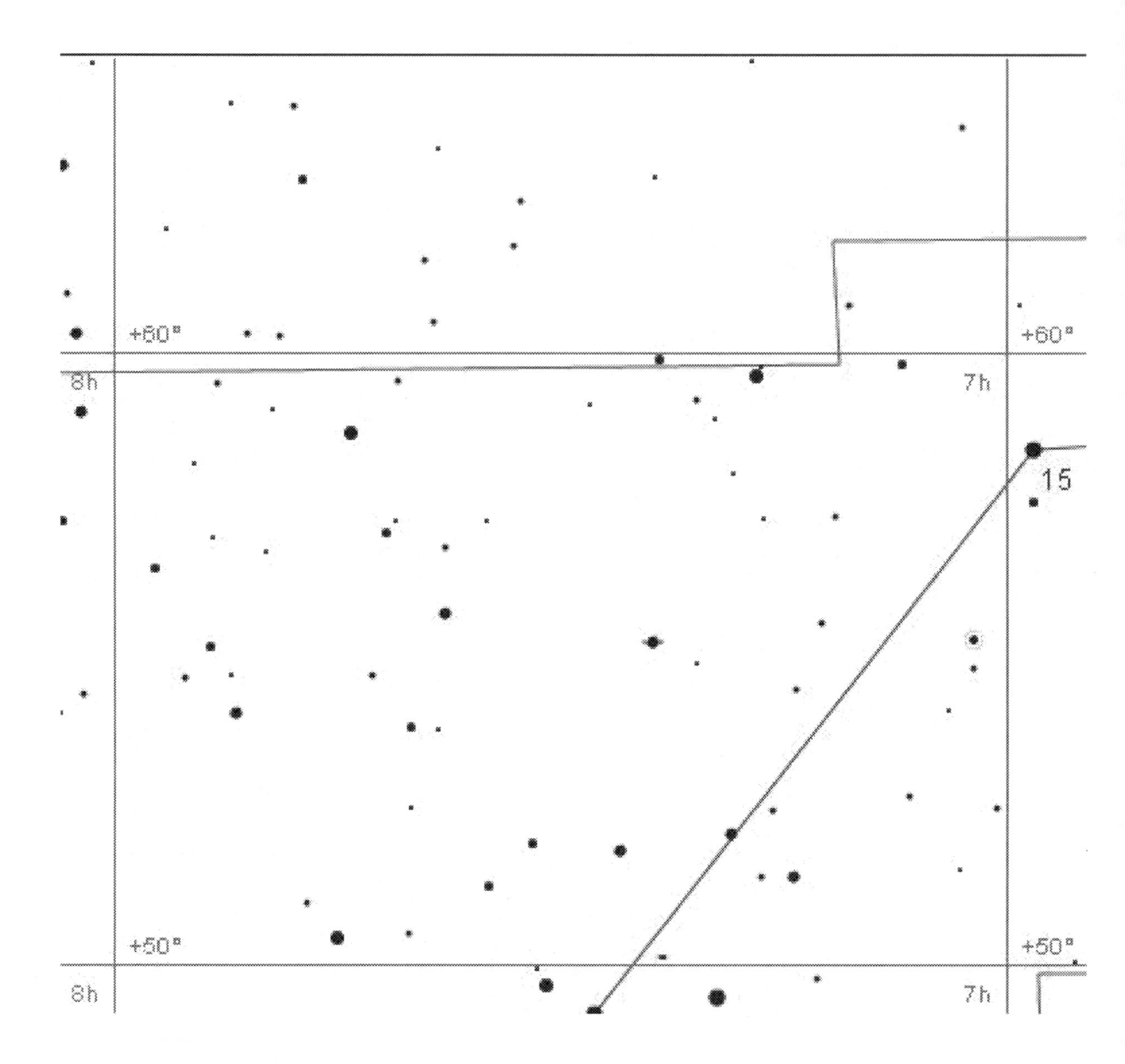
+60°
+60°
8h
7h
15
+50°
+50°
8h
7h

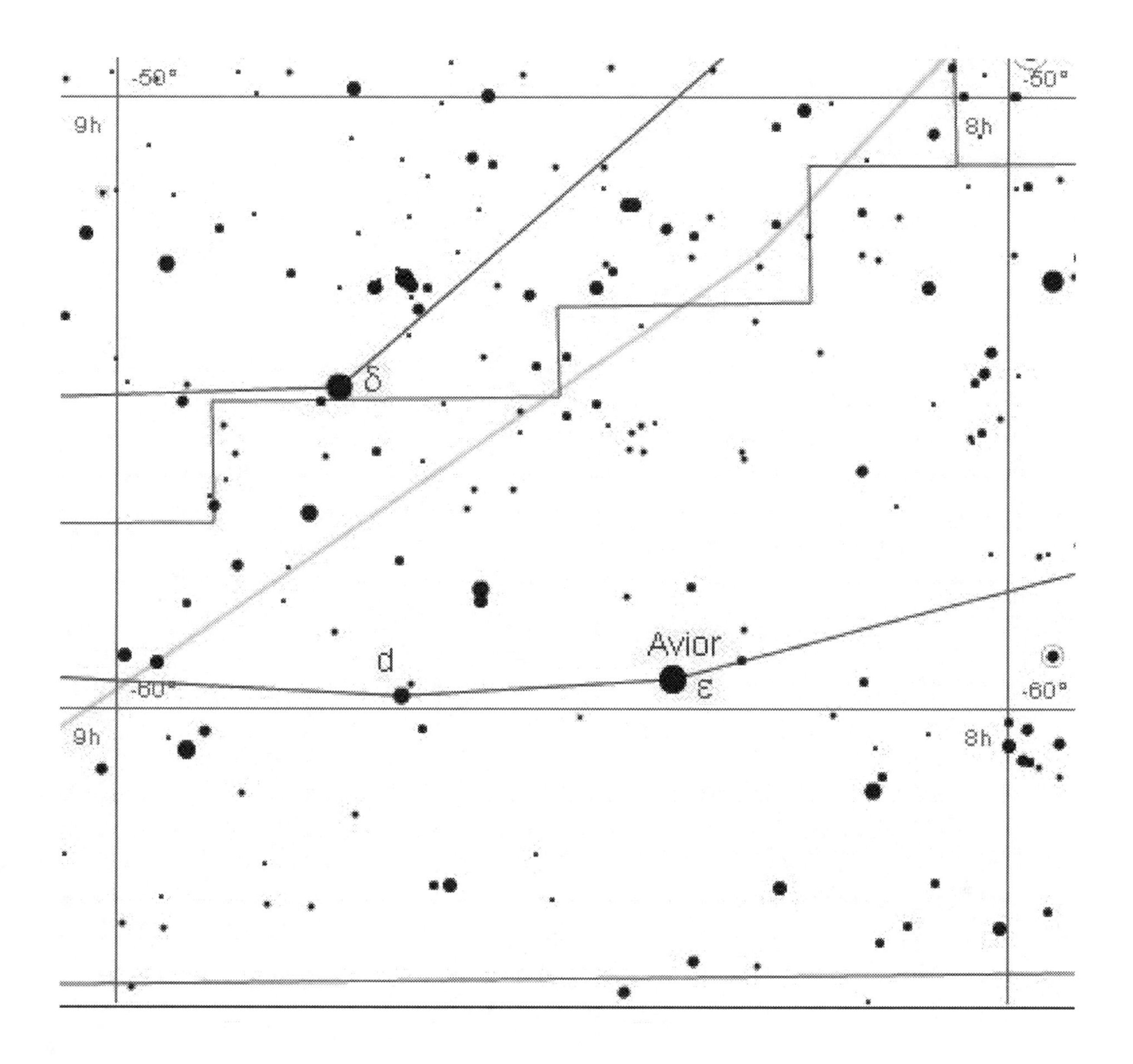

-50°
9h
8h
-50°
δ
Avior
d
ε
-60°
-60°
9h
8h

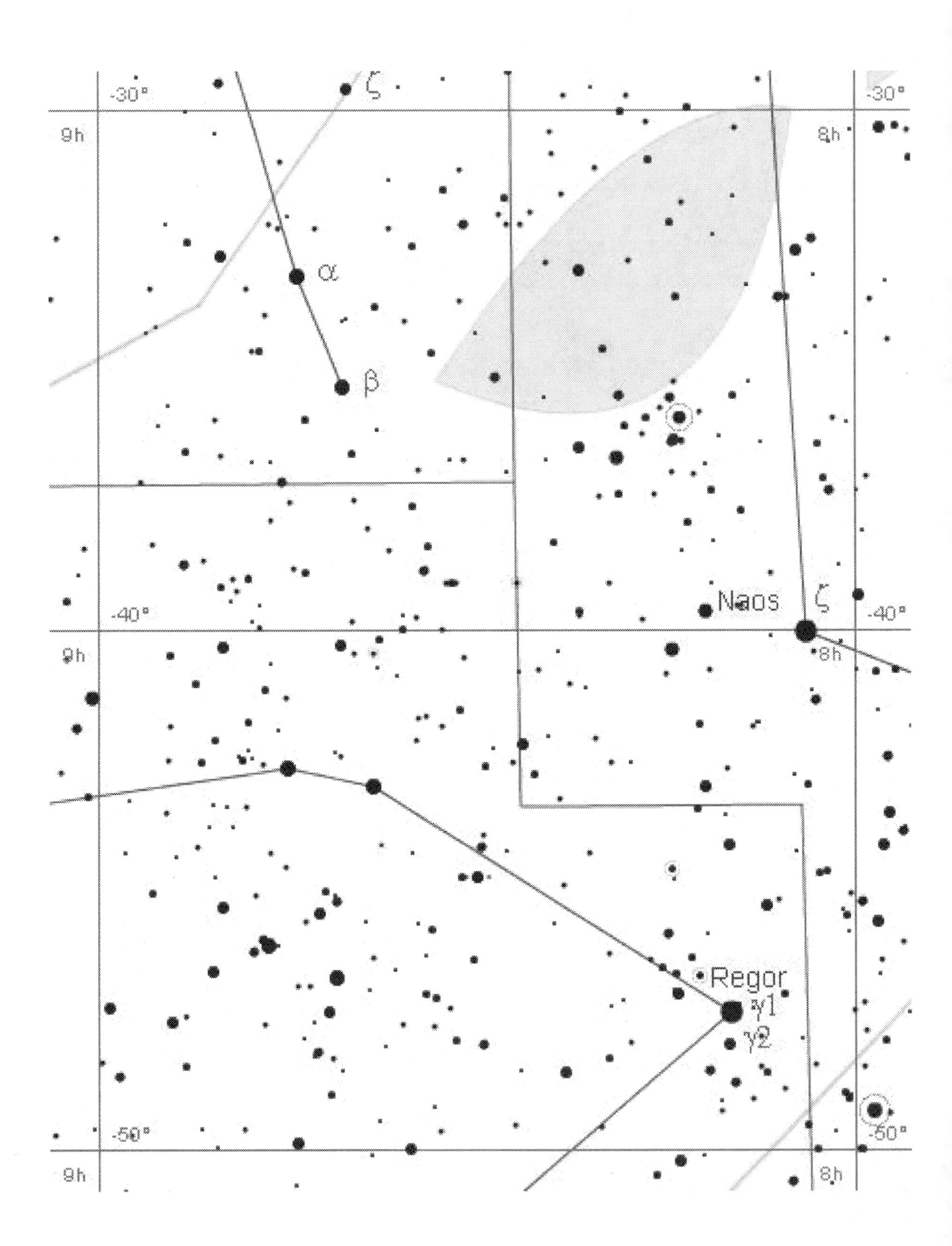
-30°
9h
ζ
α
β
8h
-30°
Naos
ζ
-40°
9h
-40°
8h
Regor
γ1
γ2
-50°
9h
-50°
8h

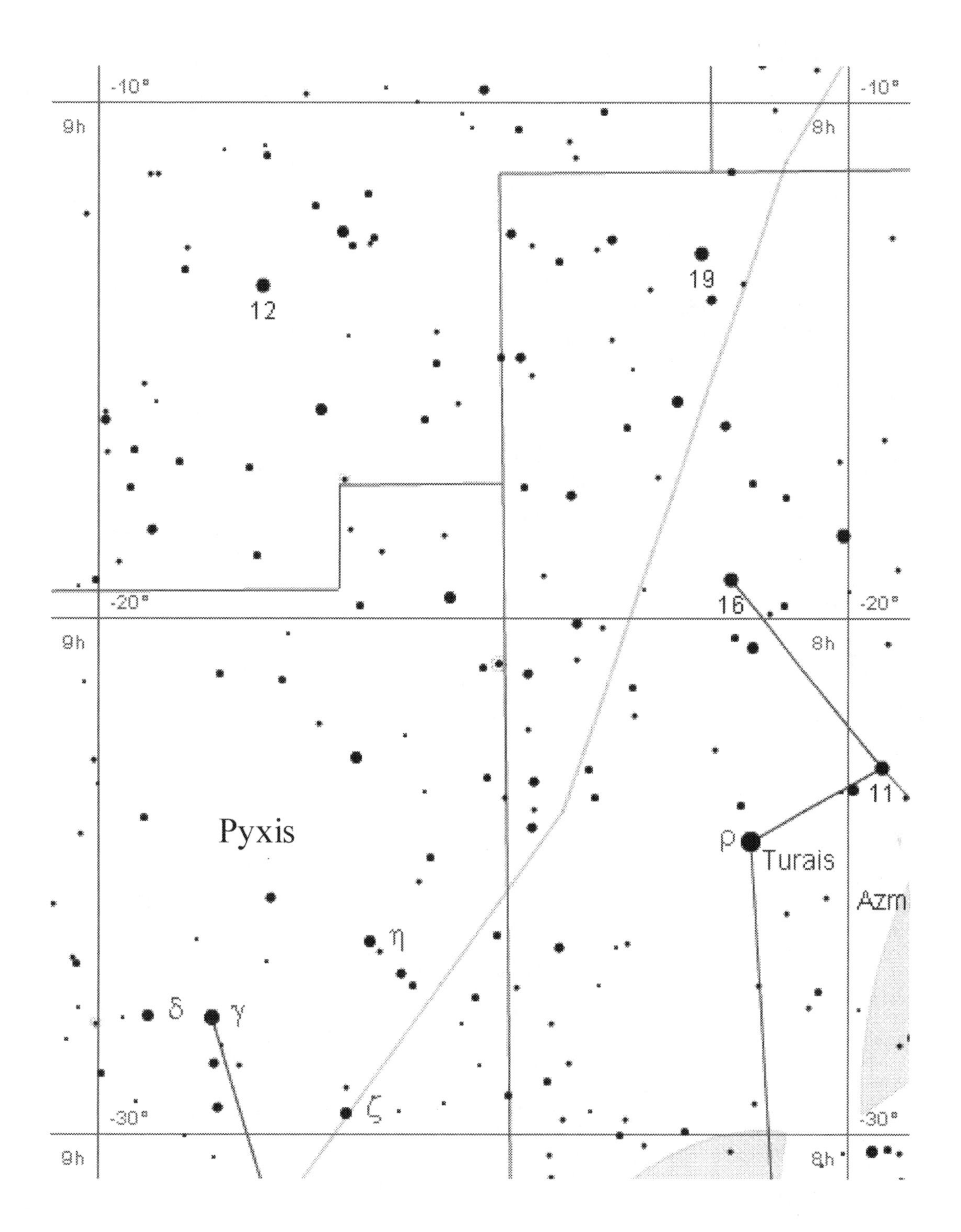
-10°
9h
8h
12
19
-20°
16
Pyxis
11
ρ
Turais
Azm
η
δ
γ
ζ
-30°

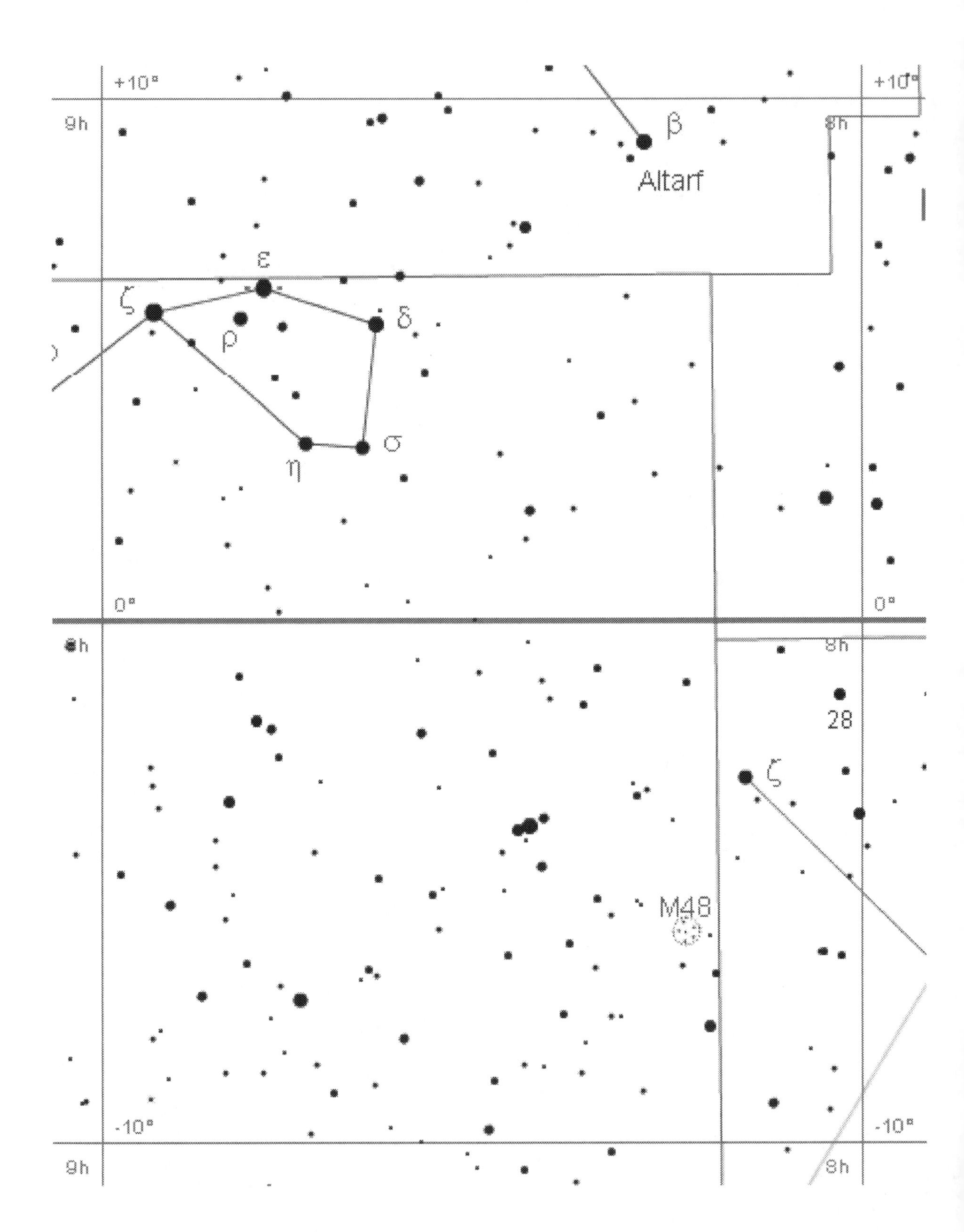
+10°
9h
β
Altarf
8h
ε
ζ
ρ
δ
σ
η
0°
9h
8h
28
ζ
M48
-10°
9h
8h

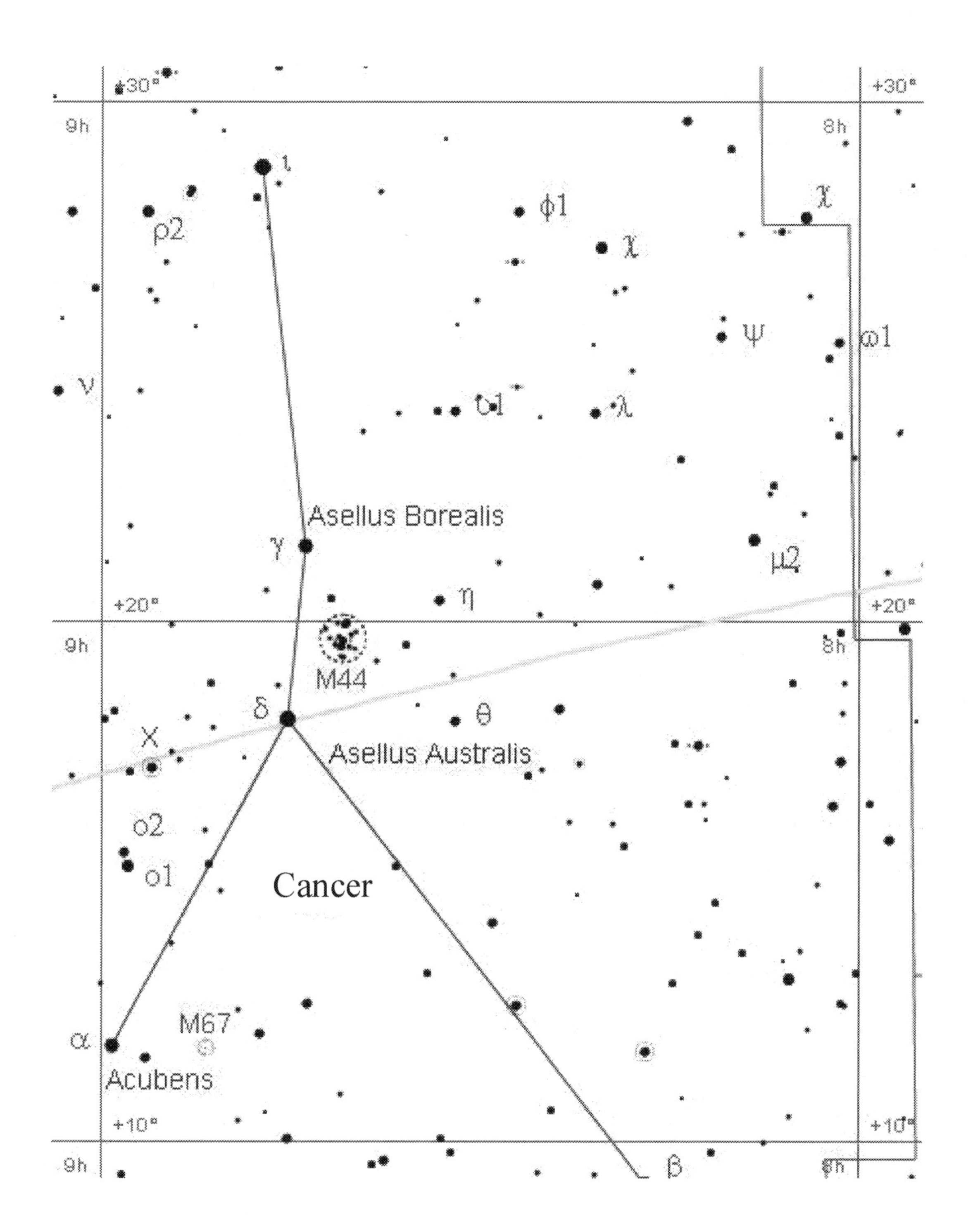
+30°
9h
8h
ι
ρ2
ϕ1
χ
ψ
ω1
ν
υ1
λ
Asellus Borealis
γ
μ2
+20°
η
M44
δ
θ
Asellus Australis
o2
o1
Cancer
M67
α
Acubens
+10°
β

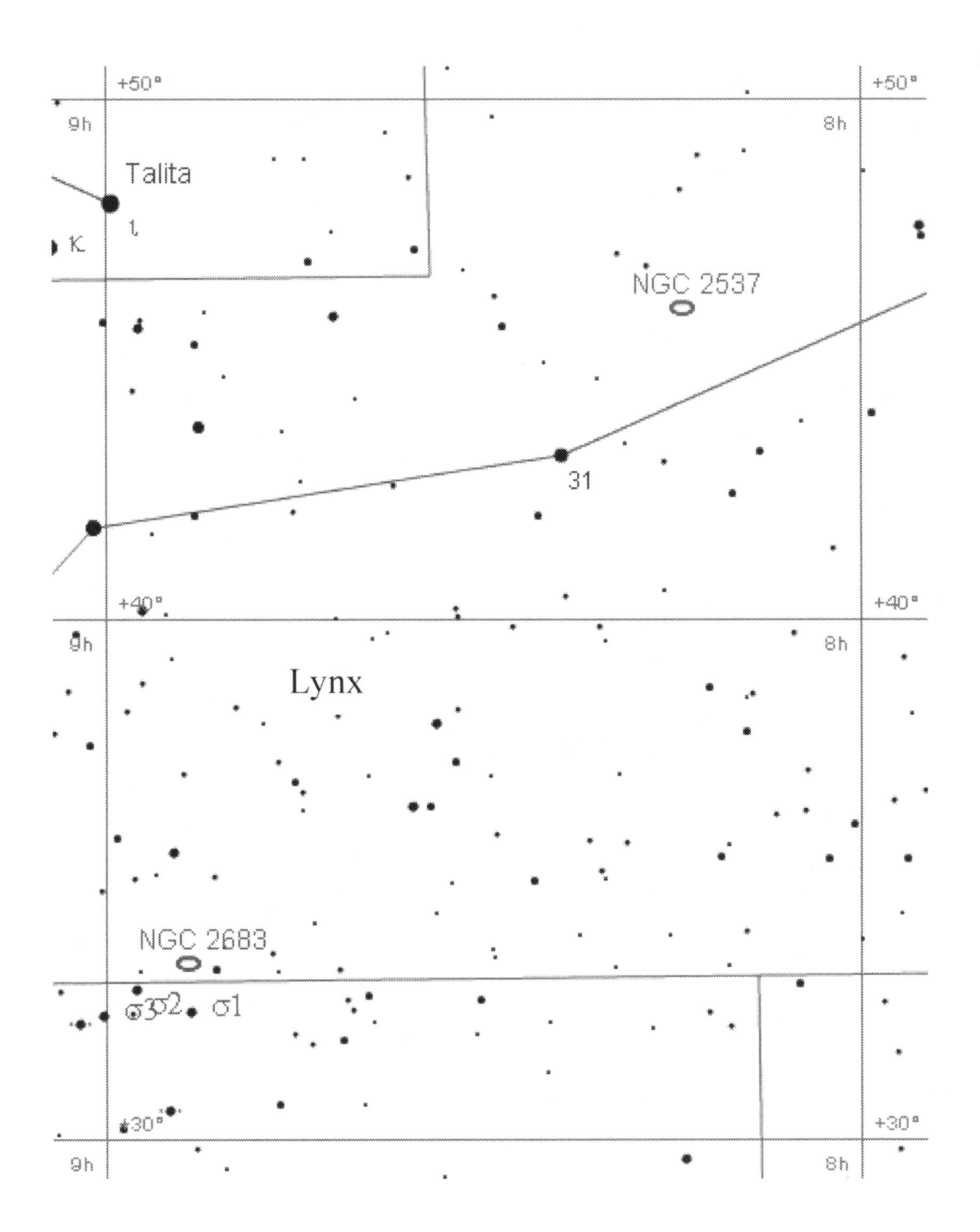
+50°
+50°
9h
8h
Talita
ι
κ
NGC 2537
31
+40°
+40°
9h
8h
Lynx
NGC 2683
σ3σ2
σ1
+30°
+30°
9h
8h

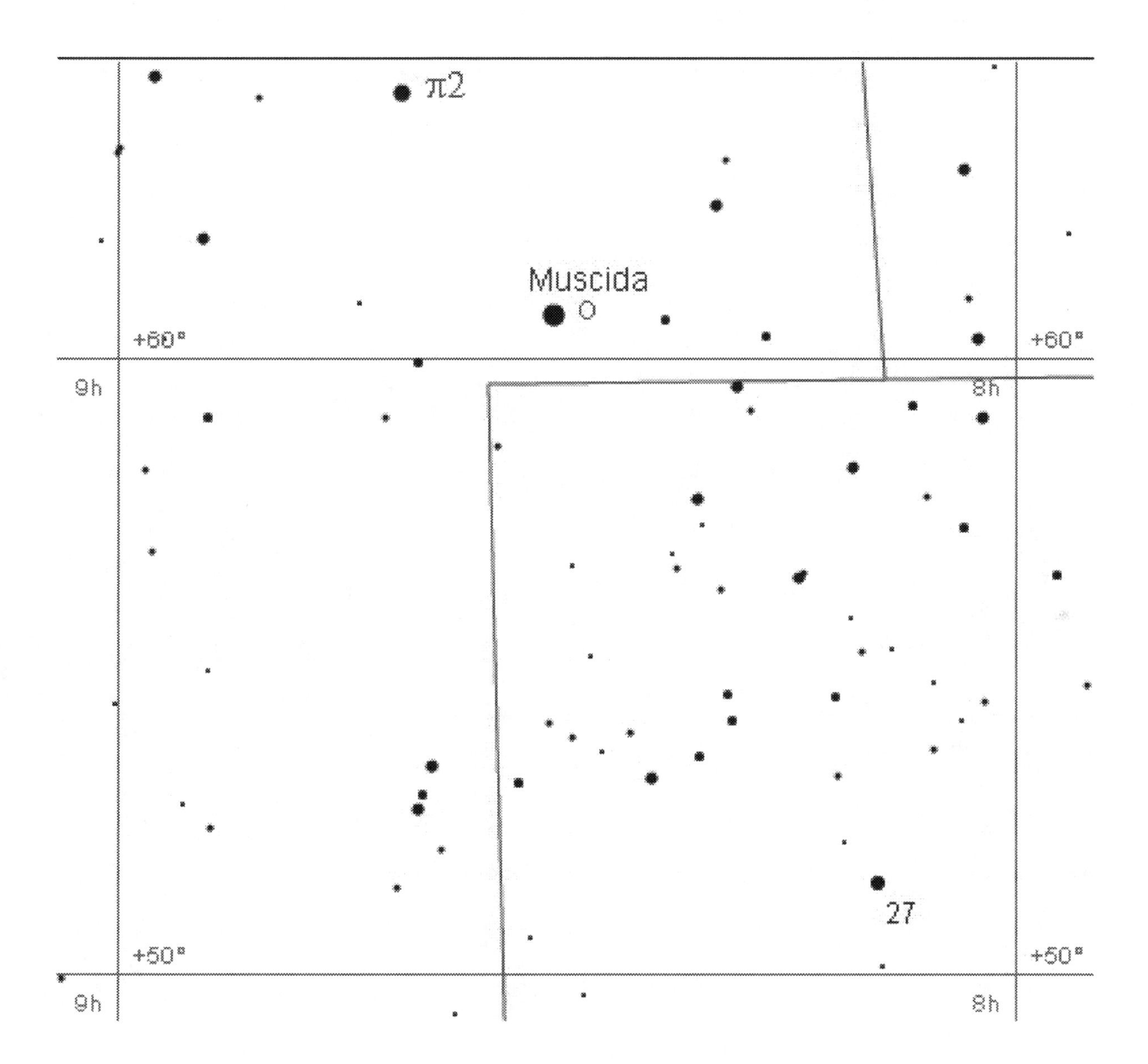
π2
Muscida
O
+60°
+60°
9h
8h
27
+50°
+50°
9h
8h

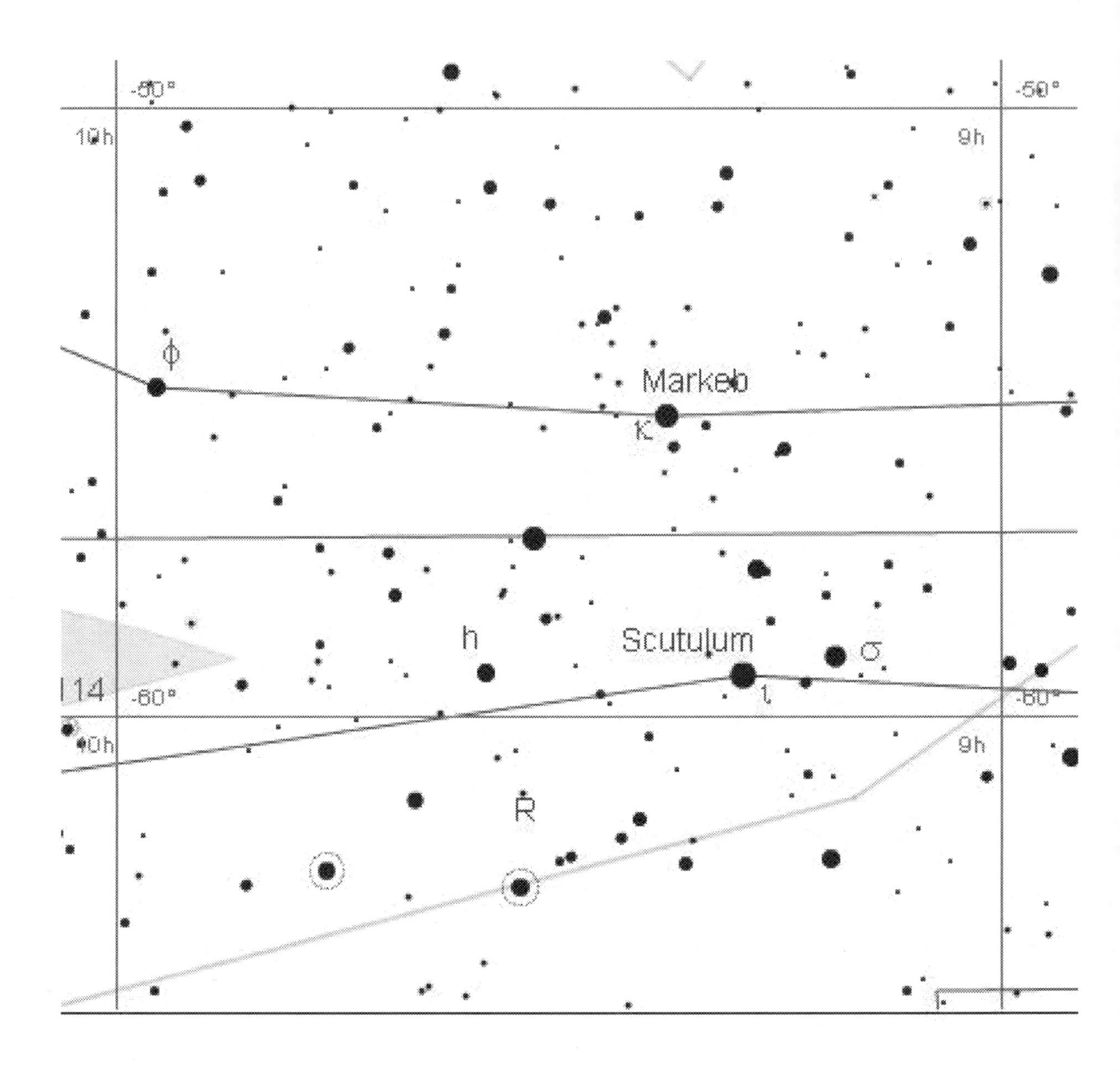
-50°
10h
9h
-50°
ϕ
Markeb
κ
h
Scutulum
σ
ι
114
-60°
-60°
10h
9h
R

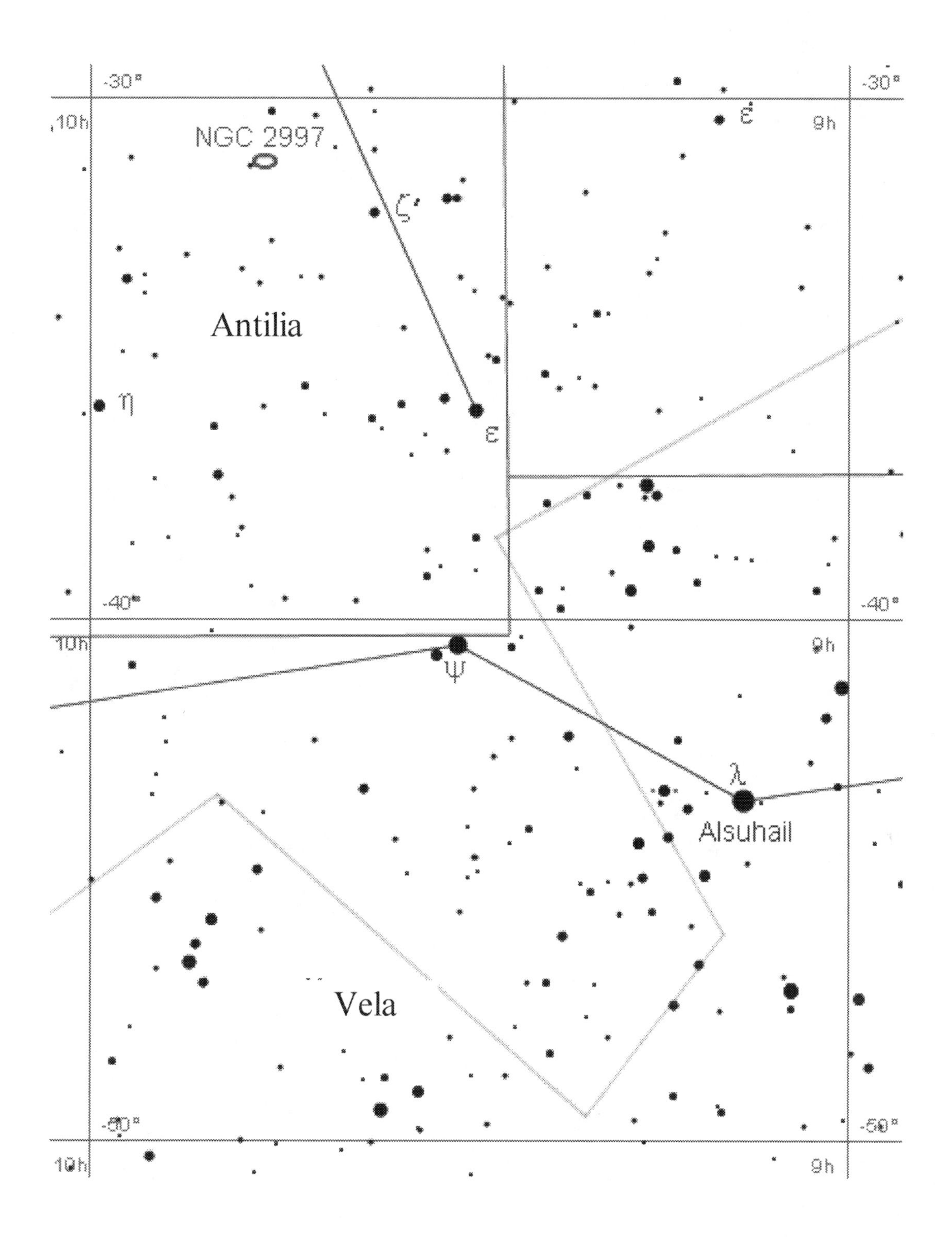
-30°
10h
NGC 2997
ζ'
ε
9h
Antilia
η
ε
-40°
10h
ψ
9h
λ
Alsuhail
Vela
-50°
10h
9h

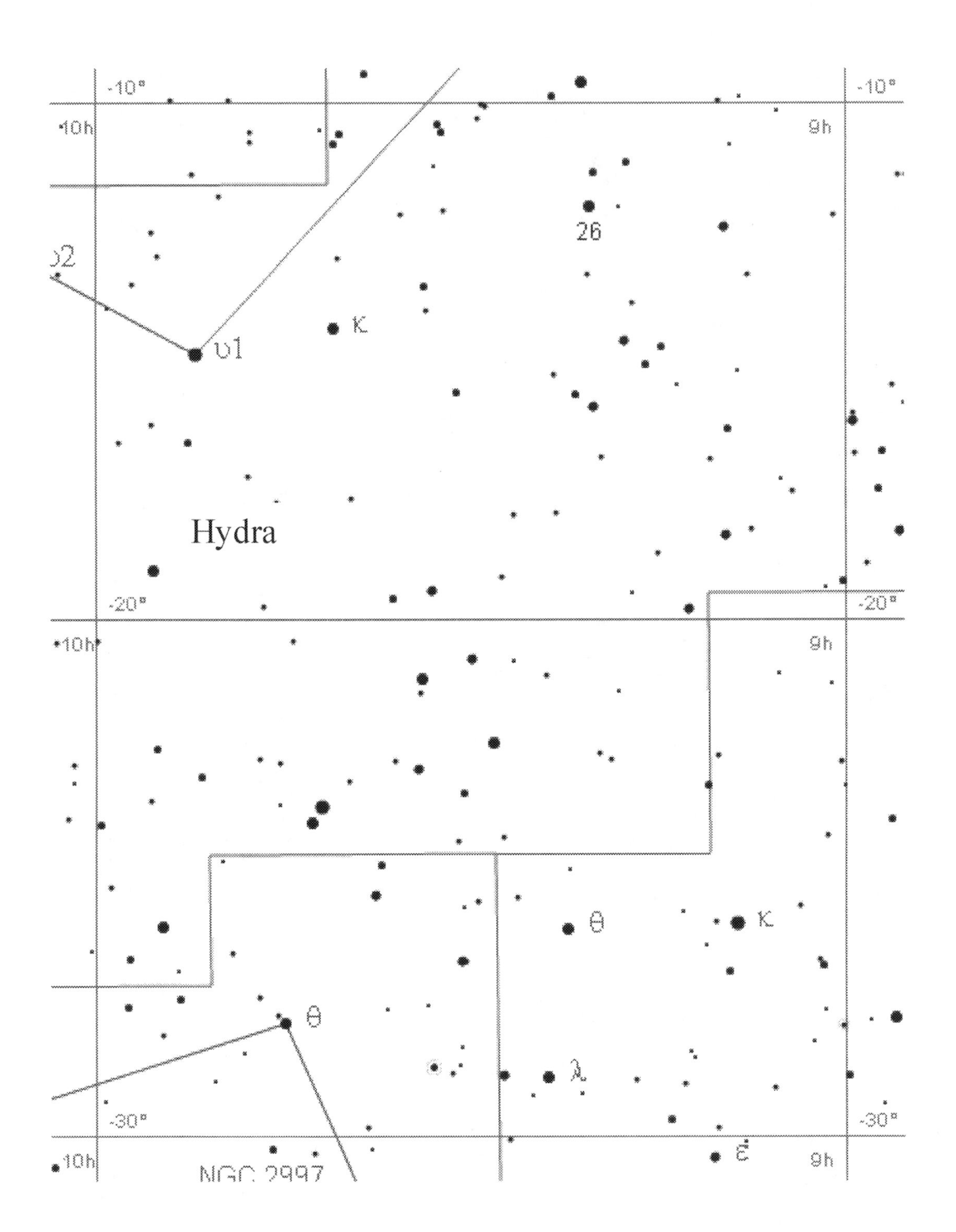
-10°
10h
9h
υ2
26
κ
υ1
Hydra
-20°
10h
9h
θ
κ
θ
λ
-30°
10h
ε
9h
NGC 2997

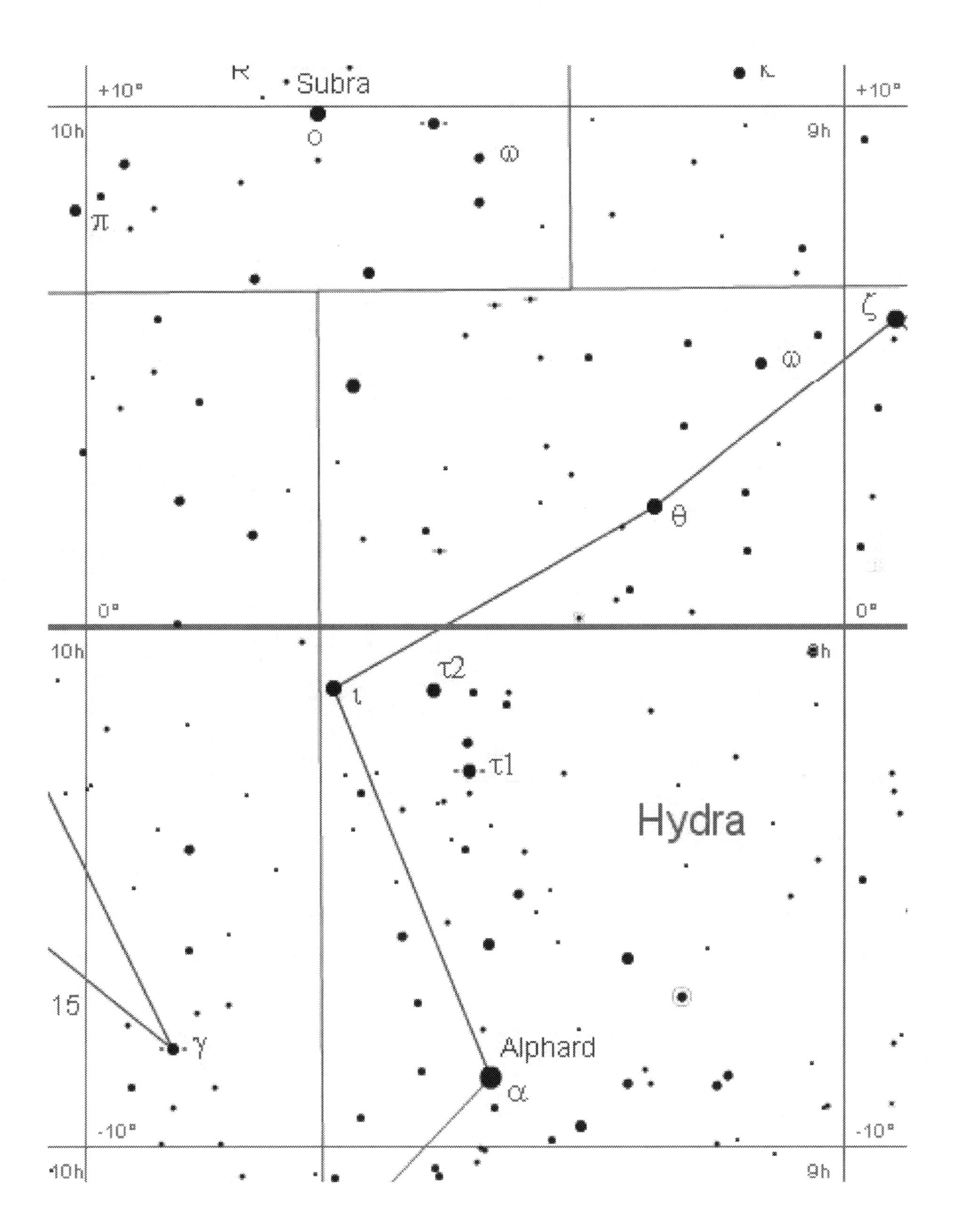
+10°
Subra
+10°
10h
9h
ζ
θ
0°
0°
10h
τ2
τ1
Hydra
15
γ
Alphard
α
-10°
-10°
10h
9h

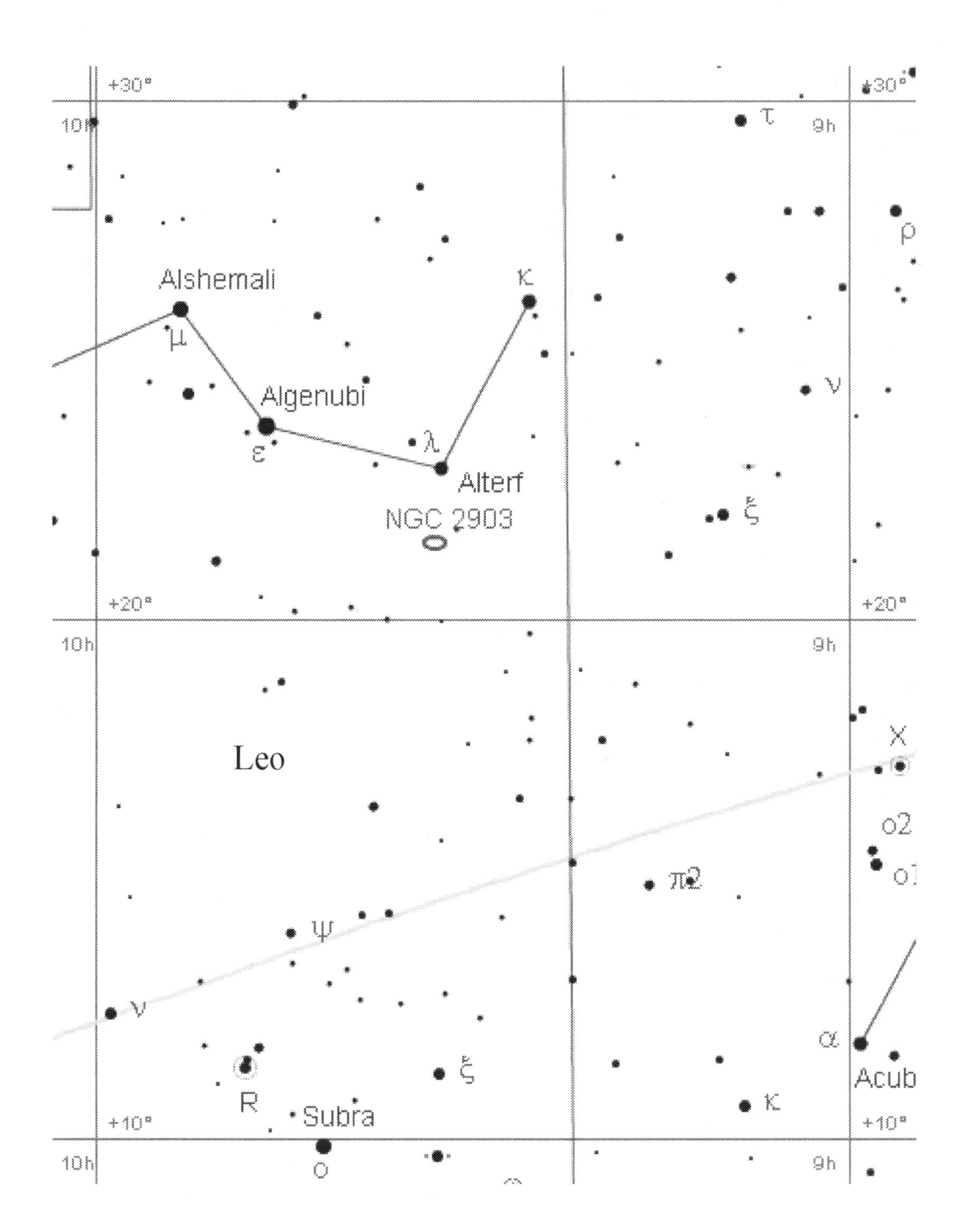
+30°
10h
9h
Alshemali
μ
Algenubi
ε
λ
Alterf
κ
NGC 2903
τ
ρ
ν
ξ
+20°
10h
9h
Leo
X
o2
π2
o1
ψ
ν
α
R
ξ
Subra
κ
Acub
+10°
10h
o
9h

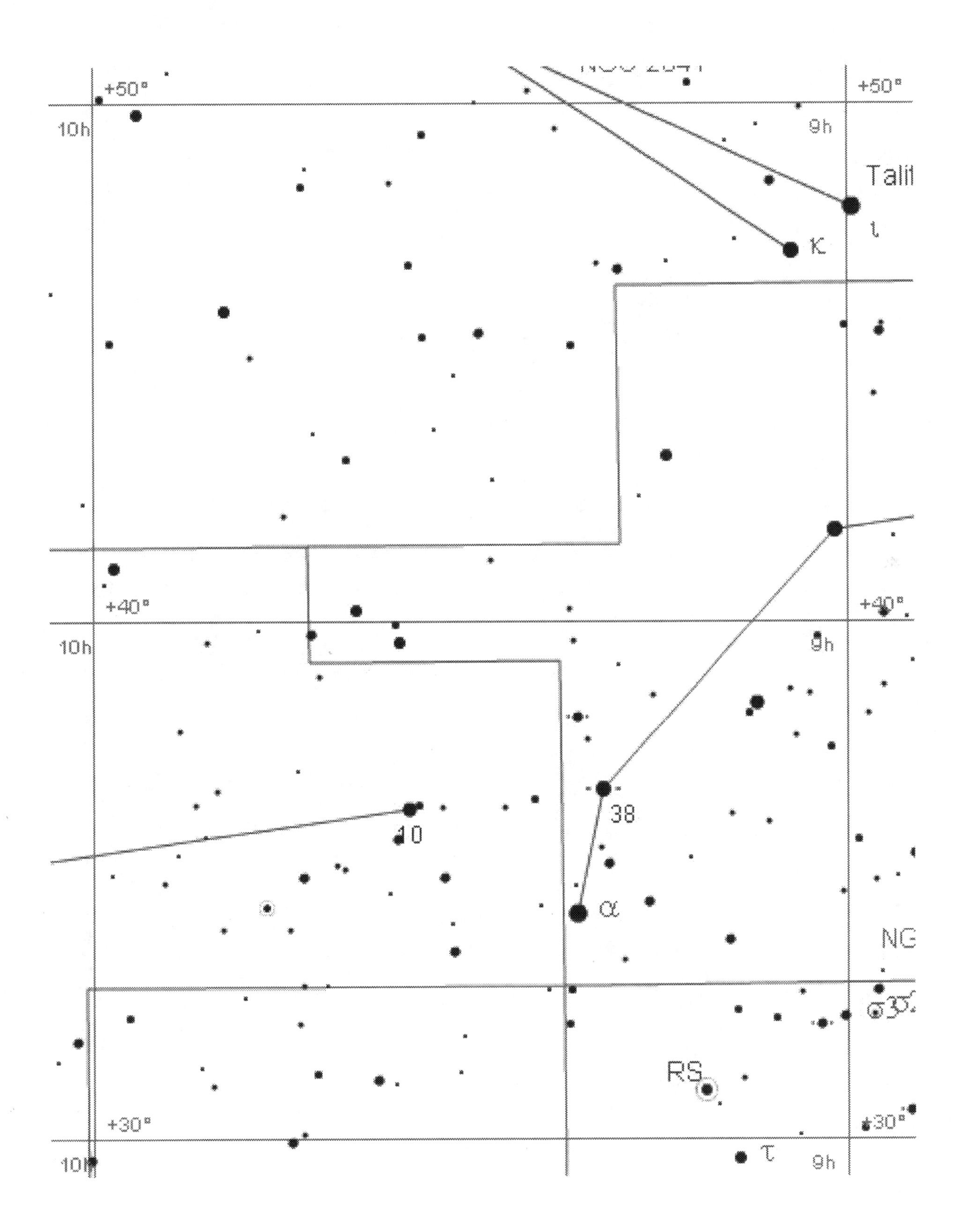
+50°
10h
9h
Talit
ι
κ
+40°
10h
9h
38
10
α
NG
RS
+30°
10h
τ
9h

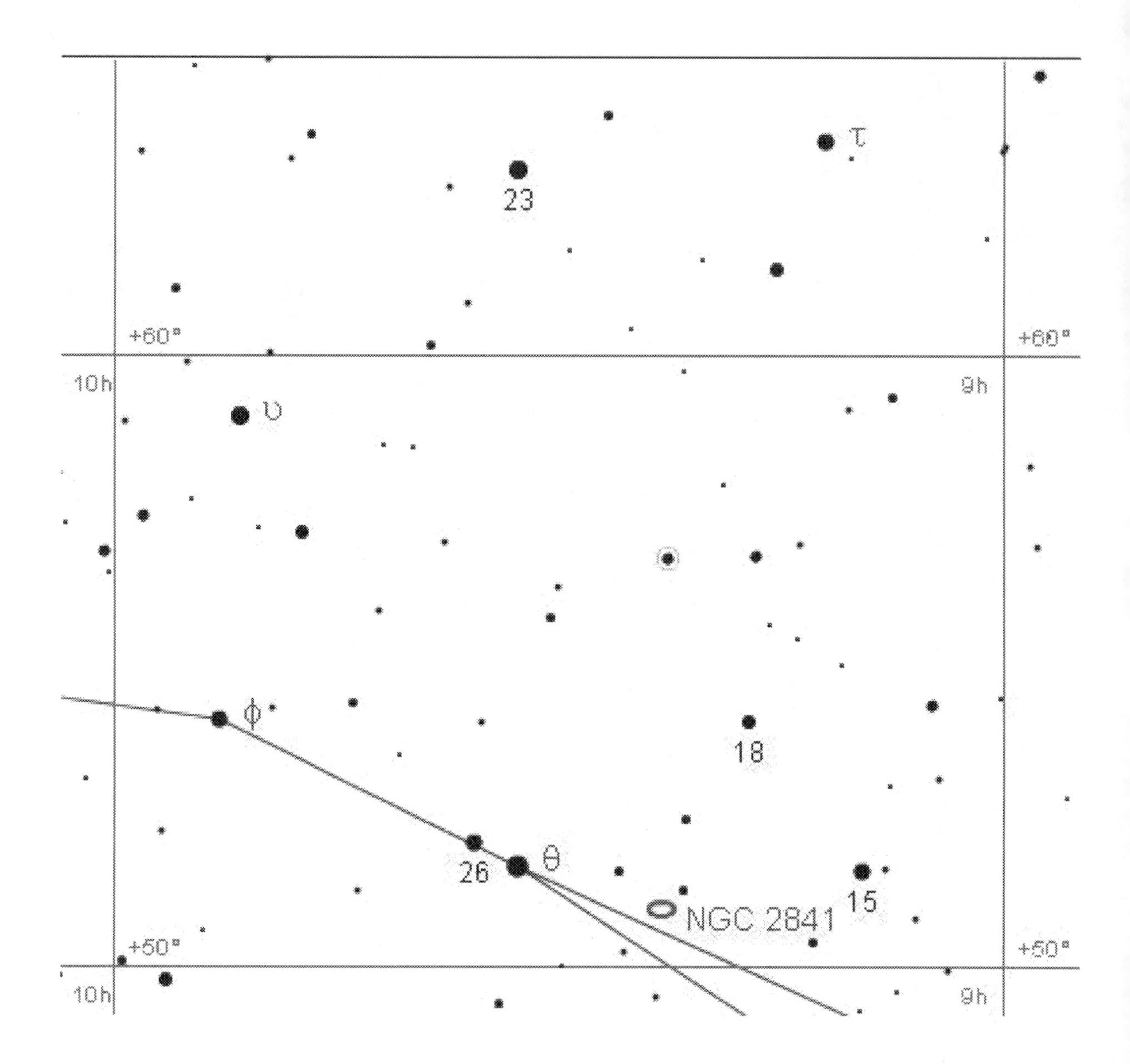
τ
23
+60°
+60°
10h
9h
υ
φ
18
26
θ
15
NGC 2841
+50°
+50°
10h
9h

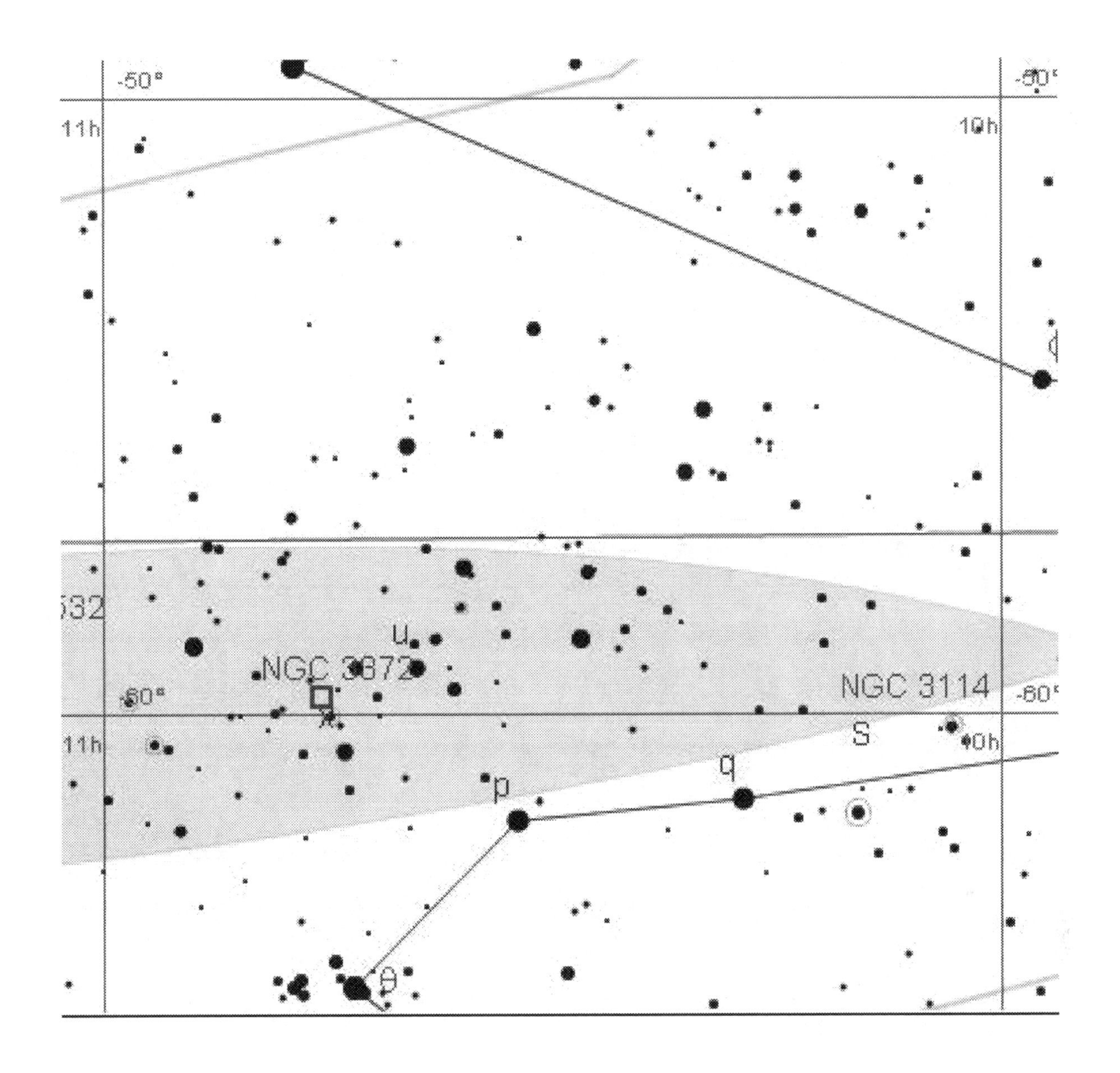
-50°
11h
10h
532
u
NGC 3372
NGC 3114
-60°
S
q
p
θ

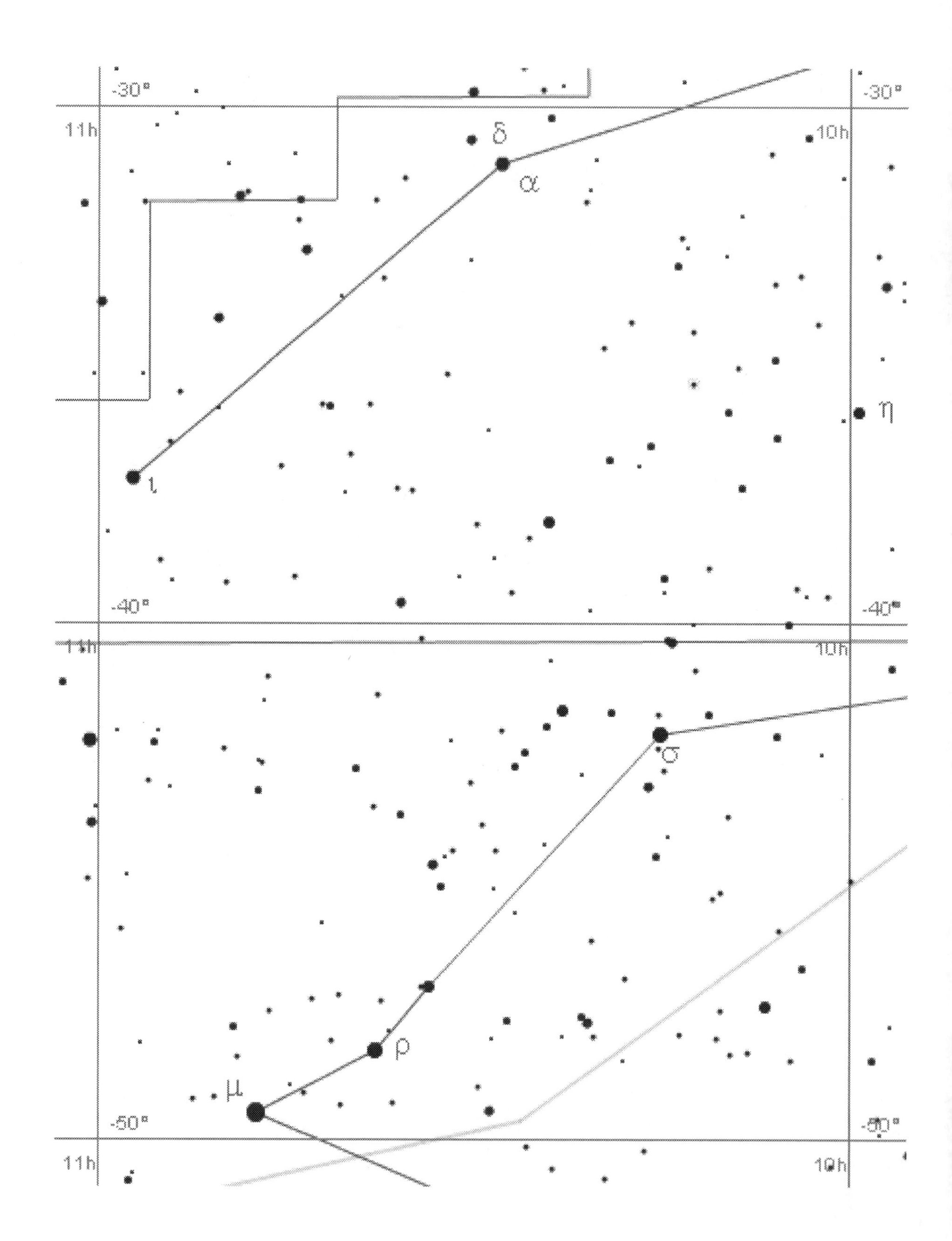

-30°
11h
δ
10h
α
η
ι
-40°
11h
10h
σ
ρ
μ
-50°
11h
10h

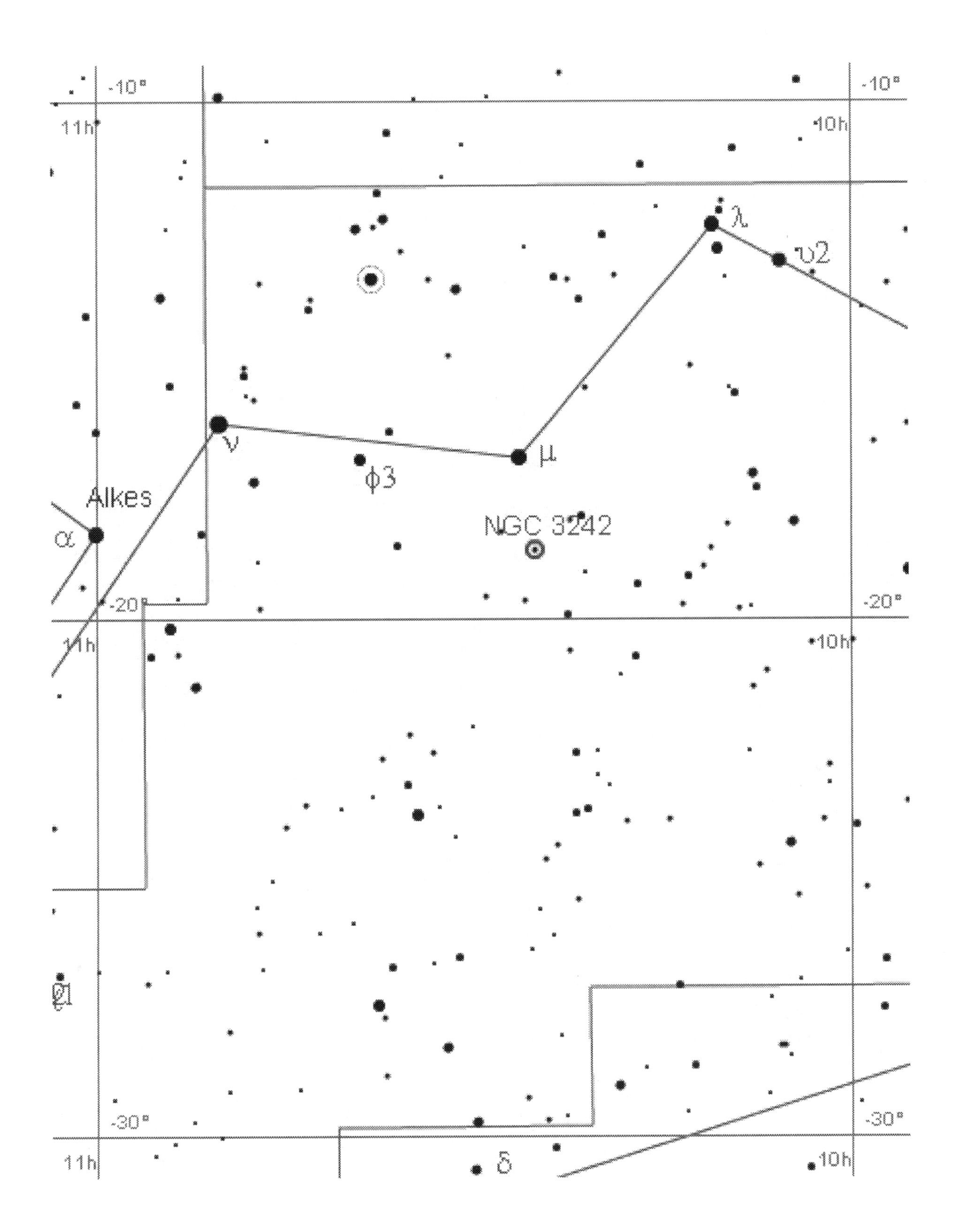
-10°
11h
10h
λ
υ2
ν
μ
φ3
Alkes
α
NGC 3242
-20°
-30°
δ

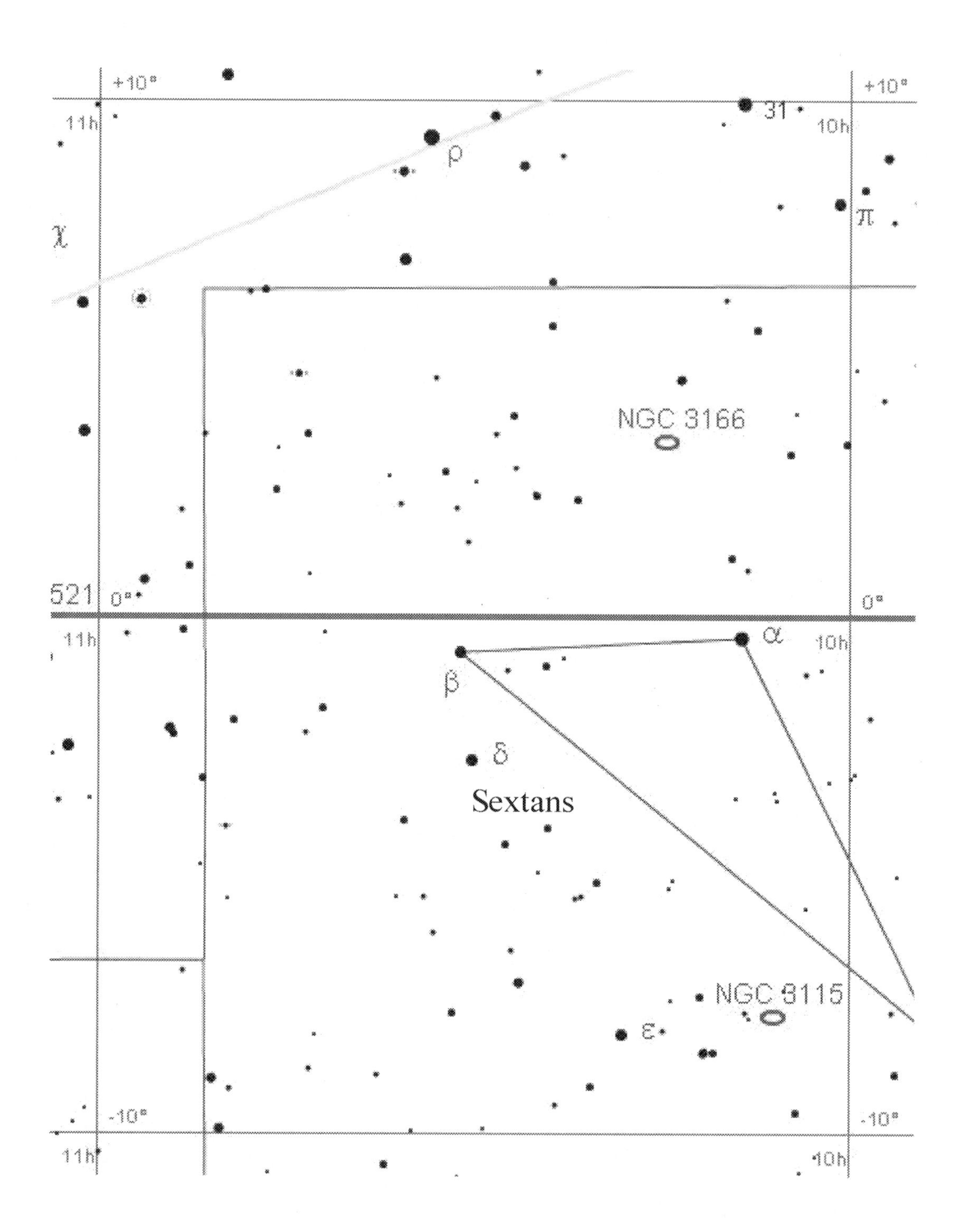
+10°
+10°
11h
10h
31
ρ
π
χ
NGC 3166
521
0°
0°
11h
10h
α
β
δ
Sextans
NGC 3115
ε
-10°
-10°
11h
10h

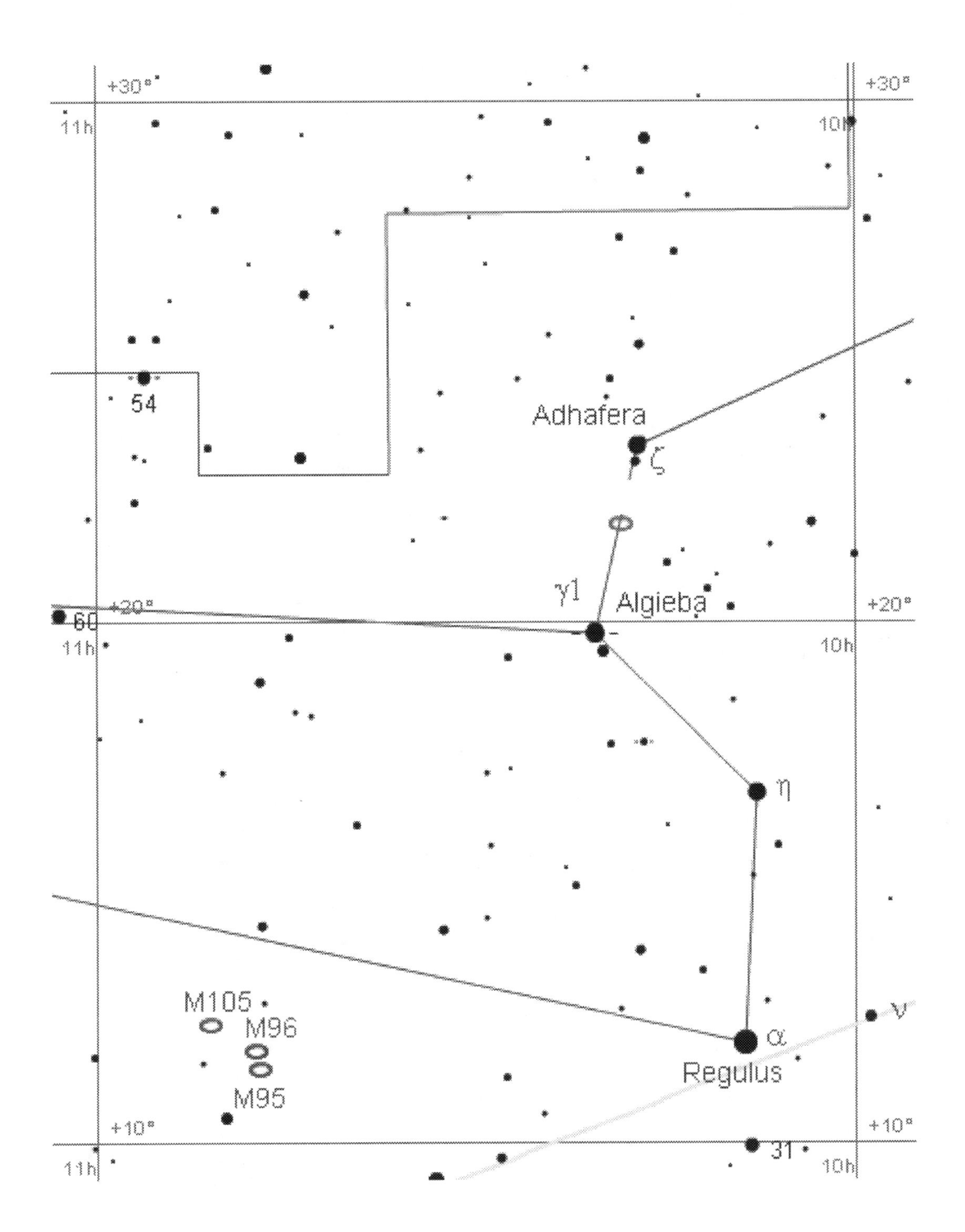

+30°
+30°
11h
10h
54
Adhafera
ζ
γ1
Algieba
+20°
+20°
60
11h
10h
η
M105
M96
α
Regulus
M95
+10°
+10°
31
11h
10h

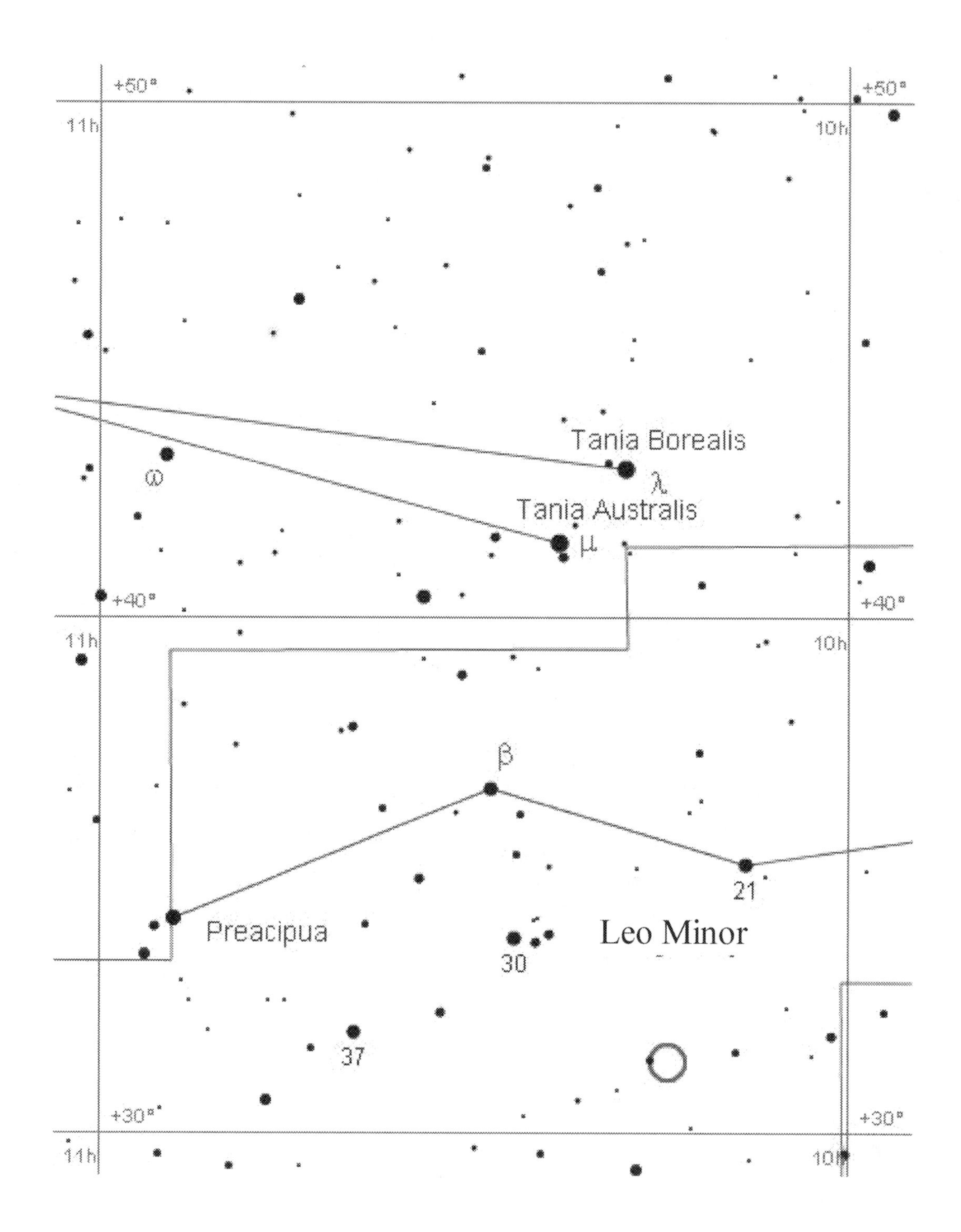
+50°
11h
10h
Tania Borealis
λ
Tania Australis
μ
ω
+40°
β
21
Preacipua
Leo Minor
30
37
+30°

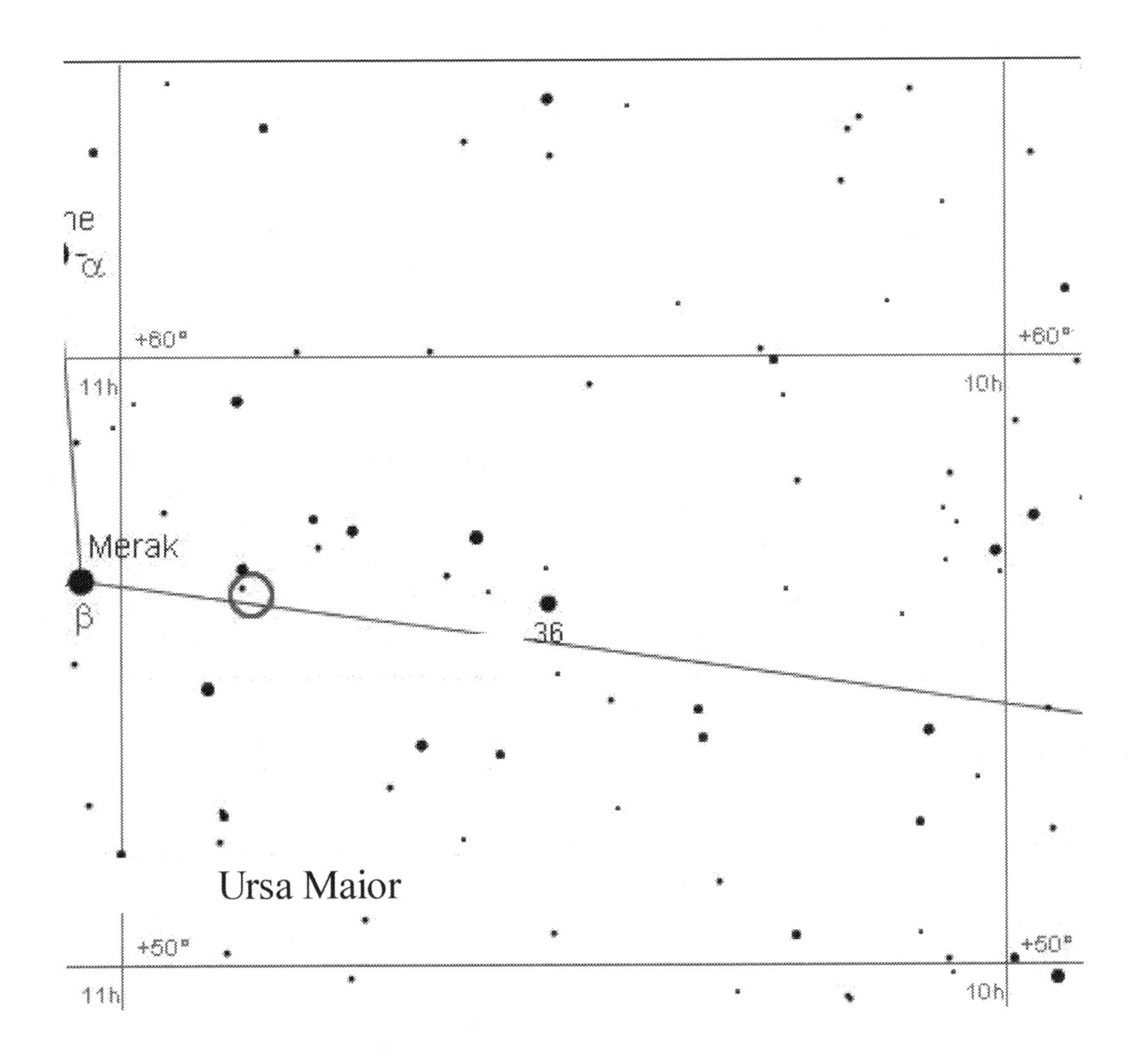
+60°
11h
10h
Merak
β
36
Ursa Maior
+50°

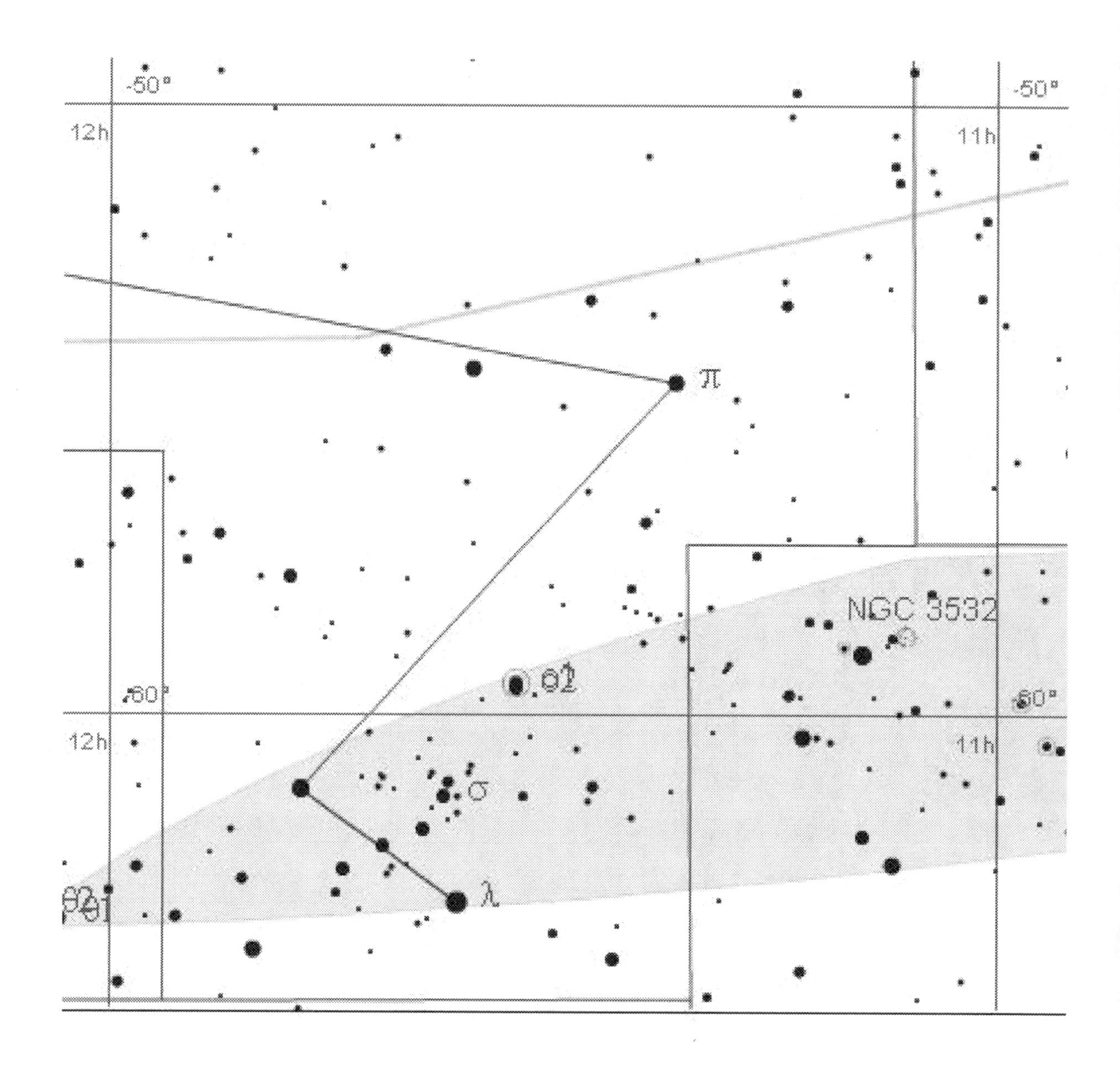
-50°
-50°
12h
11h
π
NGC 3532
-60°
-60°
12h
11h
σ
λ

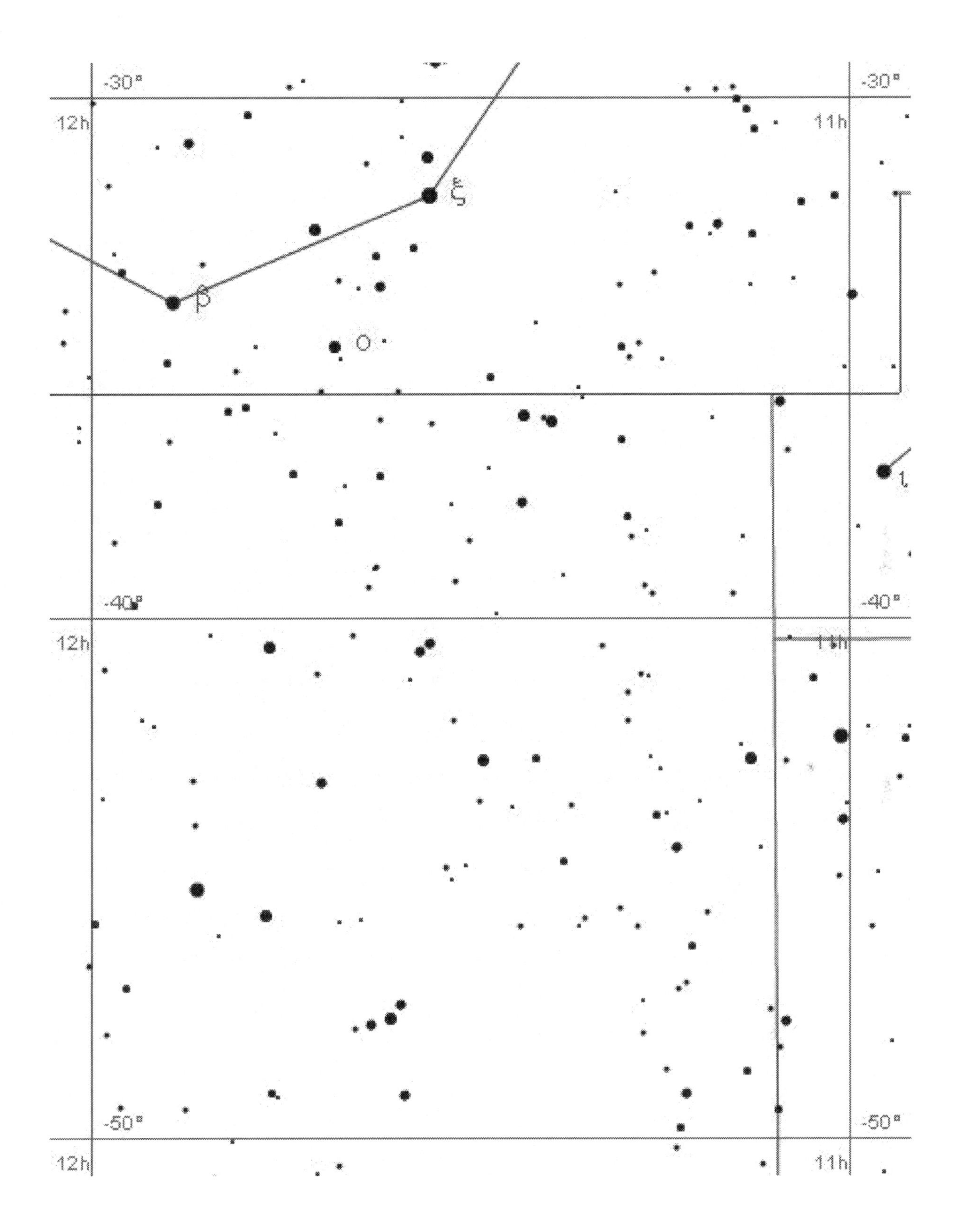
-30°
12h
11h
-30°
ξ
β
ο
-40°
-40°
12h
-50°
-50°
12h
11h

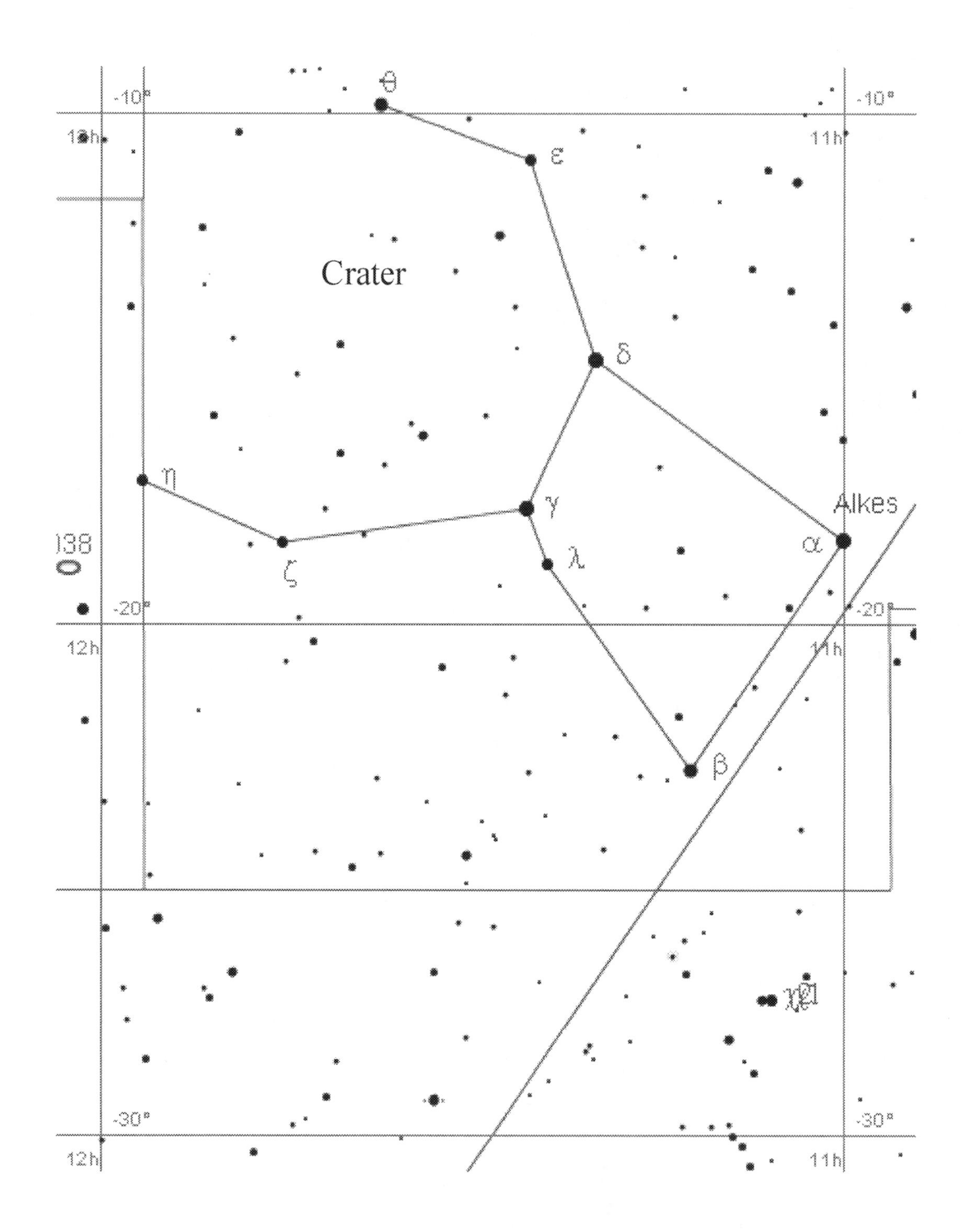
Crater
θ
ε
δ
γ
λ
η
ζ
β
α
Alkes
-10°
-20°
-30°
12h
11h

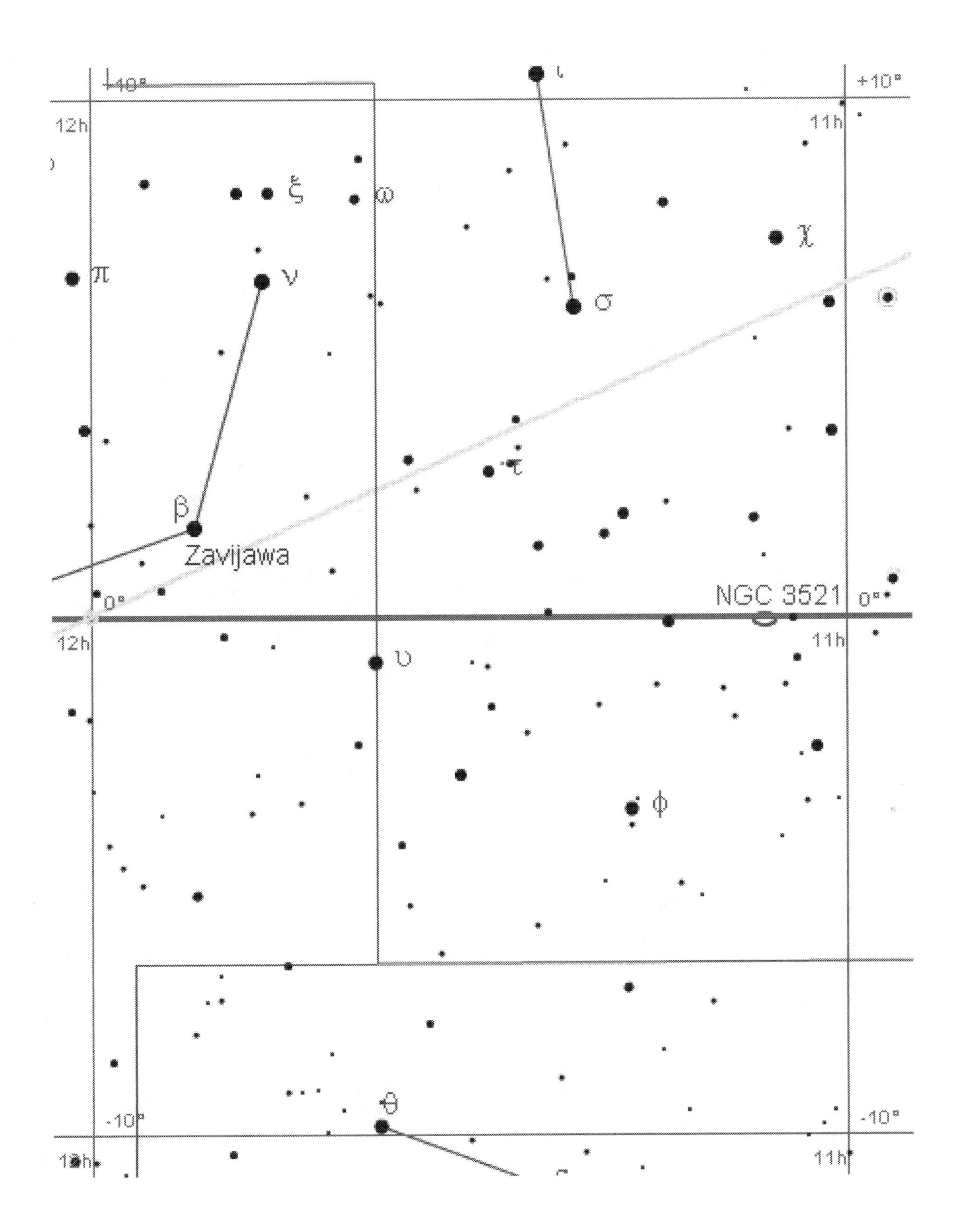
+10°
+10°
12h
11h
ξ
ω
χ
π
ν
σ
τ
β
Zavijawa
NGC 3521
0°
0°
12h
11h
υ
ϕ
θ
-10°
-10°
12h
11h

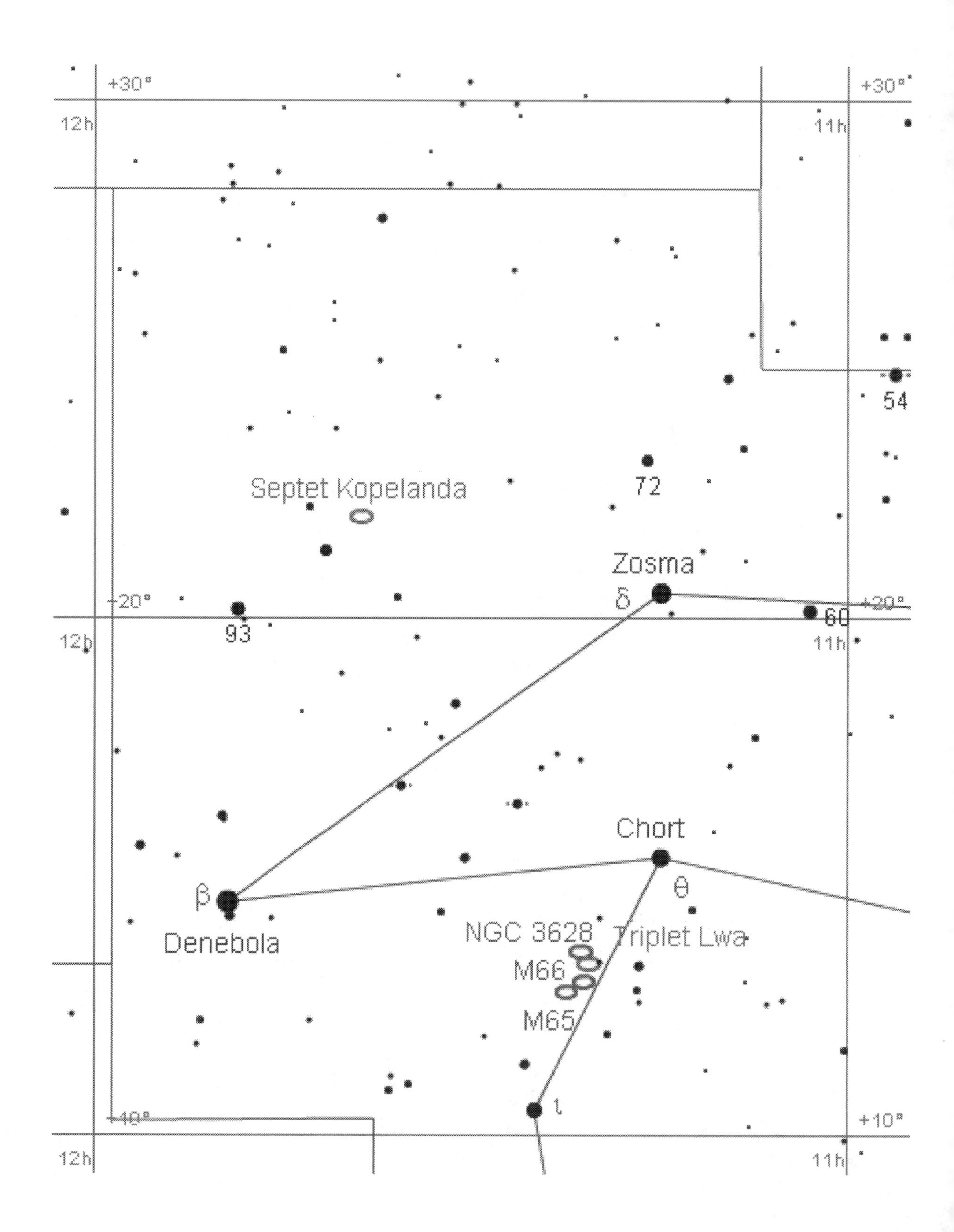
+30°
12h
11h
+30°
54
Septet Kopelanda
72
Zosma
δ
+20°
93
12h
60
11h
Chort
θ
β
Denebola
NGC 3628
Triplet Lwa
M66
M65
ι
+10°
12h
11h

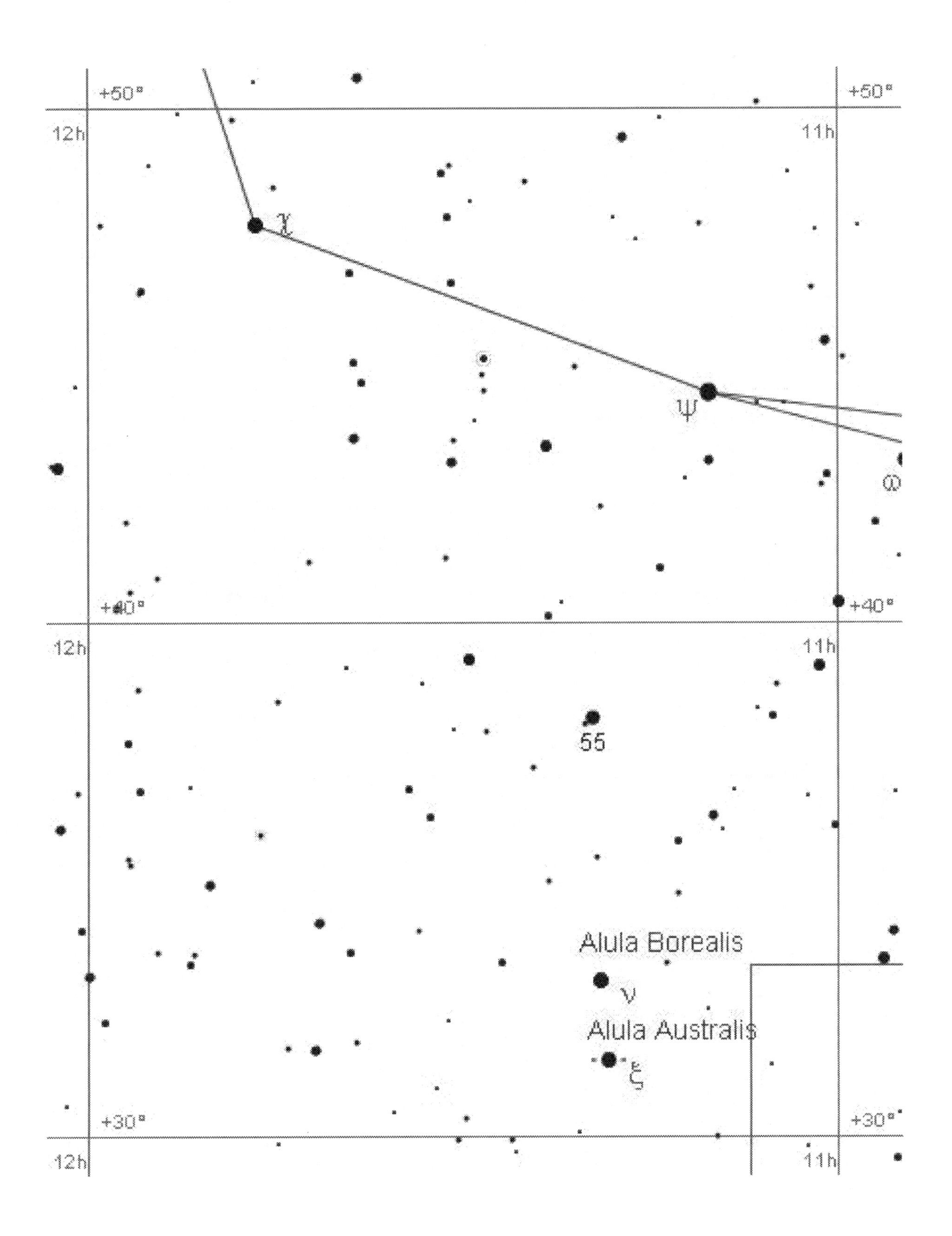
+50°
12h
11h
+50°
χ
ψ
ω
+40°
+40°
12h
11h
55
Alula Borealis
ν
Alula Australis
ξ
+30°
+30°
12h
11h

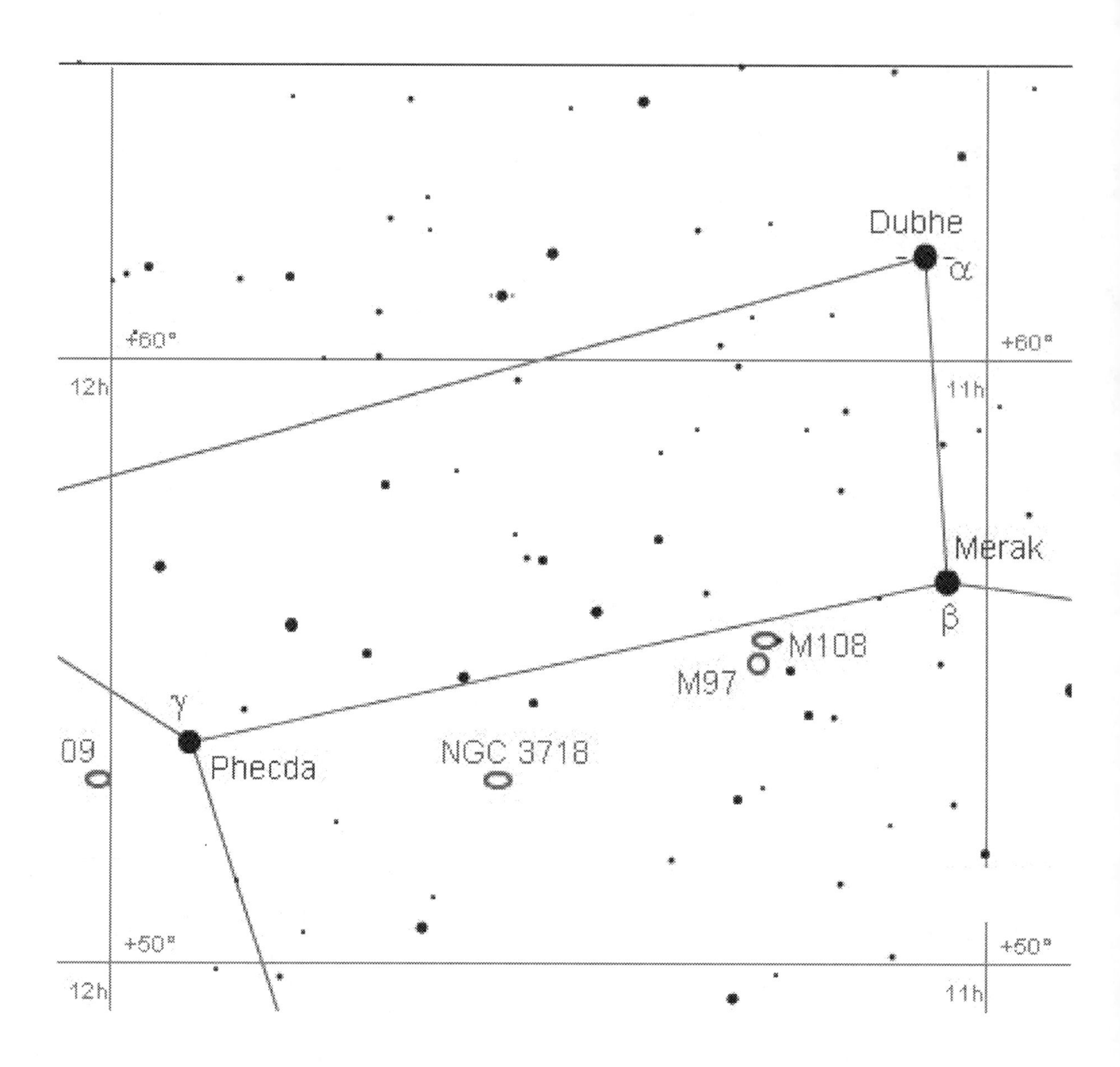
Dubhe
α
+60°
12h
11h
+60°
Merak
β
M108
M97
γ
09
Phecda
NGC 3718
+50°
12h
+50°
11h

| Catalog Name | Common Name | RA | Dec | Gal Long | Gal Lat | Spectral Type | Vis Mag | Abs Mag | Prllx |
|---|---|---|---|---|---|---|---|---|---|
| Alpha Canis Majoris | Sirius | 06 45 | -16.7 | 227.2 | -8.9 | A1V | -1.46 | 1.43 | 379.21 |
| Alpha Carinae | Canopus | 06 24 | -52.7 | 261.2 | -25.3 | F0Ib | -0.73 | -5.64 | 10.43 |
| Alpha Centauri | Rigil Kentaurus | 14 40 | -60.8 | 315.8 | -0.7 | G2V+K1V | -0.29 | 4.06 | 742.12 |
| Alpha Boötis | Arcturus | 14 16 | +19.2 | 15.2 | +69.0 | K2III | -0.05 | -0.31 | 88.85 |
| Alpha Lyrae | Vega | 18 37 | +38.8 | 67.5 | +19.2 | A0V | 0.03 | 0.58 | 128.93 |
| Alpha Aurigae | Capella | 05 17 | +46.0 | 162.6 | +4.6 | G5III | 0.07 | -0.49 | 77.29 |
| Beta Orionis | Rigel | 05 15 | -8.2 | 209.3 | -25.1 | B8Ia | 0.15v | -6.72v | 4.22 |
| Alpha Canis Minoris | Procyon | 07 39 | +5.2 | 213.7 | +13.0 | F5IV-V | 0.36 | 2.64 | 285.93 |
| Alpha Eridani | Achernar | 01 38 | -57.2 | 290.7 | -58.8 | B3V | 0.45 | -2.77 | 22.68 |
| Alpha Orionis | Betelgeuse | 05 55 | +7.4 | 199.8 | -9.0 | M2Ib | 0.55v | -5.04v | 7.63 |
| Beta Centauri | Hadar | 14 04 | -60.4 | 311.8 | +1.2 | B1III | 0.61 | -5.42 | 6.21 |
| Alpha Aquilae | Altair | 19 51 | +8.9 | 47.8 | -9.0 | A7V | 0.77 | 2.21 | 194.44 |
| Alpha Crucis | Acrux | 12 27 | -63.1 | 300.2 | -0.4 | B0.5IV+B1V | 0.79 | -4.17 | 10.17 |
| Alpha Tauri | Aldebaran | 04 36 | +16.5 | 181.0 | -20.2 | K5III | 0.86v | -0.64v | 50.09 |
| Alpha Scorpii | Antares | 16 29 | -26.4 | 351.9 | +15.1 | M1Ib+B4V | 0.95v | -5.39v | 5.40 |
| Alpha Virginis | Spica | 13 25 | -11.2 | 316.1 | +50.8 | B1V+B2V | 0.97 | -3.56 | 12.44 |
| Beta Geminorum | Pollux | 07 45 | +28.0 | 192.2 | +23.3 | K0III | 1.14 | 1.07 | 96.74 |
| Alpha Piscis Austrini | Fomalhaut | 22 58 | -29.6 | 20.6 | -65.0 | A3V | 1.15 | 1.72 | 130.08 |
| Alpha Cygni | Deneb | 20 41 | +45.3 | 84.3 | +2.1 | A2Ia | 1.24 | -8.74 | 1.01 |
| Beta Crucis | Mimosa | 12 48 | -59.7 | 302.5 | +3.2 | B0.5III | 1.26 | -3.91 | 9.25 |
| Alpha Leonis | Regulus | 10 08 | +12.0 | 226.3 | +48.9 | B7V | 1.36 | -0.52 | 42.09 |
| Epsilon Canis Majoris | Adhara | 06 59 | -29.0 | 239.9 | -11.3 | B2II | 1.50 | -4.10 | 7.57 |
| Alpha Geminorum | Castor | 07 35 | +31.9 | 187.5 | +22.6 | A1V+A2V | 1.58 | 0.59 | 63.27 |
| Lambda Scorpii | Shaula | 17 34 | -37.1 | 351.8 | -2.3 | B2IV | 1.62 | -5.05 | 4.64 |
| Gamma Crucis | Gacrux | 12 31 | -57.1 | 300.2 | +5.7 | M3.5III | 1.63 | -0.52 | 37.09 |
| Gamma Orionis | Bellatrix | 05 25 | +6.3 | 197.0 | -16.0 | B2III | 1.64 | -2.72 | 13.42 |
| Beta Tauri | Elnath | 05 26 | +28.6 | 178.0 | -3.8 | B7III | 1.66 | -1.36 | 24.89 |
| Beta Carinae | Miaplacidus | 09 13 | -69.7 | 286.0 | -14.4 | A2III | 1.67 | -0.99 | 29.34 |
| Epsilon Orionis | Alnilam | 05 36 | -1.2 | 205.2 | -17.3 | B0Ia | 1.69 | -6.38 | 2.43 |
| Alpha Gruis | Alnair | 22 08 | -47.0 | 350.0 | -52.4 | B7IV | 1.74 | -0.72 | 32.16 |
| Zeta Orionis | Alnitak | 05 41 | -1.9 | 206.5 | -16.5 | O9.5Ib+B0III | 1.75 | -5.25 | 3.99 |
| Epsilon Ursae Majoris | Alioth | 12 54 | +56.0 | 122.2 | +61.1 | A0IV | 1.77 | -0.20 | 40.30 |
| Alpha Persei | Mirfak | 03 24 | +49.9 | 146.5 | -5.9 | F5Ib | 1.80 | -4.49 | 5.51 |
| Alpha Ursae Majoris | Dubhe | 11 04 | +61.8 | 142.8 | +51.0 | K0III+F0V | 1.80 | -1.09 | 26.38 |
| Gamma Velorum | Regor | 08 10 | -47.3 | 262.8 | -7.6 | WC8+O9Ib | 1.81 | -5.25 | 3.88 |
| Delta Canis Majoris | Wezen | 07 08 | -26.4 | 238.4 | -8.3 | F8Ia | 1.83 | -6.87 | 1.82 |
| Epsilon Sagittarii | Kaus Australis | 18 24 | -34.4 | 359.2 | -9.8 | B9.5III | 1.84 | -1.39 | 22.55 |
| Eta Ursae Majoris | Alkaid | 13 48 | +49.3 | 100.5 | +65.3 | B3V | 1.86 | -0.59 | 32.39 |
| Theta Scorpii | Sargas | 17 37 | -43.0 | 347.1 | -5.9 | F1II | 1.86 | -2.75 | 11.99 |
| Epsilon Carinae | Avior | 08 23 | -59.5 | 274.3 | -12.5 | K3II+B2V | 1.87 | -4.57 | 5.16 |
| Beta Aurigae | Menkalinan | 06 00 | +44.9 | 167.5 | +10.5 | A2IV | 1.90 | -0.10 | 39.72 |
| Alpha Trianguli Australis | Atria | 16 49 | -69.0 | 321.6 | -15.3 | K2Ib-II | 1.92 | -3.61 | 7.85 |
| Gamma Geminorum | Alhena | 06 38 | +16.4 | 196.8 | +4.5 | A0IV | 1.93 | -0.60 | 31.12 |
| Alpha Pavonis | Peacock | 20 26 | -56.7 | 340.9 | -35.3 | B0.5V+B2V | 1.93 | -1.82 | 17.80 |
| Delta Velorum | Koo She | 08 45 | -54.7 | 272.1 | -7.3 | A0V | 1.95 | 0.01 | 40.90 |
| Beta Canis Majoris | Mirzam | 06 23 | -18.0 | 226.1 | -14.2 | B1III | 1.98 | -3.95 | 6.53 |
| Alpha Hydrae | Alphard | 09 28 | -8.7 | 241.6 | +29.1 | K3II | 1.98 | -1.70 | 18.40 |
| Alpha Ursae Minoris | Polaris | 02 32 | +89.3 | 123.3 | +26.5 | F7Ib-II | 1.99v | -3.62v | 7.56 |
| Gamma Leonis | Algieba | 10 20 | +19.8 | 216.6 | +54.7 | K0III+G7III | 2.00 | -0.93 | 25.96 |
| Alpha Arietis | Hamal | 02 07 | +23.5 | 144.5 | -36.2 | K2III | 2.01 | 0.48 | 49.48 |
| Beta Ceti | Diphda | 00 44 | -18.0 | 112.0 | -80.7 | K0III | 2.04 | -0.30 | 34.04 |
| Sigma Sagittarii | Nunki | 18 55 | -26.3 | 9.5 | -12.4 | B3V | 2.05 | -2.14 | 14.54 |
| Theta Centauri | Menkent | 14 07 | -36.4 | 319.5 | +24.0 | K0III | 2.06 | 0.70 | 53.52 |
| Alpha Andromedae | Alpheratz | 00 08 | +29.1 | 111.6 | -32.8 | B9IV | 2.07 | -0.30 | 33.60 |
| Beta Andromedae | Mirach | 01 10 | +35.6 | 127.2 | -27.1 | M0II | 2.07 | -1.86 | 16.36 |
| Kappa Orionis | Saiph | 05 48 | -9.7 | 214.6 | -18.4 | B0.5III | 2.07 | -4.65 | 4.52 |

| | | | | | | | | | |
|---|---|---|---|---|---|---|---|---|---|
| Beta Ursae Minoris | Kochab | 14 51 | +74.2 | 112.7 | +40.5 | K4III | 2.07 | -0.87 | 25.79 |
| Beta Gruis | Al Dhanab | 22 43 | -46.9 | 346.2 | -58.0 | M5III | 2.07v | -1.52v | 19.17 |
| Alpha Ophiuchi | Rasalhague | 17 35 | +12.6 | 35.9 | +22.6 | A5III-IV | 2.08 | 1.30 | 69.84 |
| Beta Persei | Algol | 03 08 | +41.0 | 148.9 | -14.9 | B8V+G5IV+A | 2.09e | -0.18 | 35.14 |
| Gamma Andromedae | Almach | 02 04 | +42.3 | 137.0 | -18.6 | K3II+B8V+A0V | 2.10 | -3.08 | 9.19 |
| Beta Leonis | Denebola | 11 49 | +14.6 | 250.6 | +70.8 | A3V | 2.14 | 1.92 | 90.16 |
| Gamma Cassiopeiae | Cih | 00 57 | +60.7 | 123.6 | -2.2 | B0IV | 2.15v | -4.22v | 5.32 |
| Gamma Centauri | Muhlifain | 12 42 | -49.0 | 301.3 | +13.8 | A0III+A0III | 2.20 | -0.81 | 25.01 |
| Zeta Puppis | Naos | 08 04 | -40.0 | 256.0 | -4.6 | O5Ia | 2.21 | -5.95 | 2.33 |
| Iota Carinae | Aspidiske | 09 17 | -59.3 | 278.5 | -7.0 | A8Ib | 2.21v | -4.42v | 4.71 |
| Alpha Coronae Borealis | Alphecca | 15 35 | +26.7 | 41.9 | +53.7 | A0V+G5V | 2.22e | 0.42 | 43.65 |
| Lambda Velorum | Suhail | 09 08 | -43.4 | 265.9 | +2.9 | K4Ib | 2.23 | -3.99 | 5.69 |
| Zeta Ursae Majoris | Mizar | 13 24 | +54.9 | 113.1 | +61.6 | A2V+A2V+A1V | 2.23 | 0.33 | 41.73 |
| Gamma Cygni | Sadr | 20 22 | +40.3 | 78.2 | +1.9 | F8Ib | 2.23 | -6.12 | 2.14 |
| Alpha Cassiopeiae | Schedar | 00 41 | +56.5 | 121.5 | -6.3 | K0II | 2.24 | -1.99 | 14.27 |
| Gamma Draconis | Eltanin | 17 57 | +51.5 | 79.1 | +29.1 | K5III | 2.24 | -1.04 | 22.10 |
| Delta Orionis | Mintaka | 05 32 | -0.3 | 203.9 | -17.7 | O9.5II+B2V | 2.25e | -4.99 | 3.56 |
| Beta Cassiopeiae | Caph | 00 09 | +59.2 | 117.5 | -3.2 | F2III | 2.28 | 1.17 | 59.89 |
| Epsilon Centauri | | 13 40 | -53.5 | 310.2 | +8.7 | B1III | 2.29 | -3.02 | 8.68 |
| Delta Scorpii | Dschubba | 16 00 | -22.6 | 350.1 | +22.6 | B0.5IV | 2.29 | -3.16 | 8.12 |
| Epsilon Scorpii | Wei | 16 50 | -34.3 | 348.8 | +6.6 | K2.5III | 2.29 | 0.78 | 49.85 |
| Alpha Lupi | Men | 14 42 | -47.4 | 321.6 | +11.4 | B1.5III | 2.30 | -3.83 | 5.95 |
| Eta Centauri | | 14 36 | -42.2 | 322.9 | +16.6 | B1.5V | 2.33v | -2.55v | 10.57 |
| Beta Ursae Majoris | Merak | 11 02 | +56.4 | 149.1 | +54.8 | A1V | 2.34 | 0.41 | 41.07 |
| Epsilon Boötis | Izar | 14 45 | +27.1 | 39.4 | +64.8 | K0II-III+A2V | 2.35 | -1.69 | 15.55 |
| Epsilon Pegasi | Enif | 21 44 | +9.9 | 65.6 | -31.4 | K2Ib | 2.38 | -4.19 | 4.85 |
| Kappa Scorpii | Girtab | 17 42 | -39.0 | 351.0 | -4.6 | B1.5III | 2.39 | -3.38 | 7.03 |
| Alpha Phoenicis | Ankaa | 00 26 | -42.3 | 320.2 | -74.0 | K0III | 2.40 | 0.52 | 42.14 |
| Gamma Ursae Majoris | Phecda | 11 54 | +53.7 | 140.8 | +61.4 | A0V | 2.41 | 0.36 | 38.99 |
| Eta Ophiuchi | Sabik | 17 10 | -15.7 | 6.7 | +14.1 | A1V+A3V | 2.43 | 0.37 | 38.77 |
| Beta Pegasi | Scheat | 23 04 | +28.1 | 95.8 | -29.1 | M2III | 2.44v | -1.49v | 16.37 |
| Eta Canis Majoris | Aludra | 07 24 | -29.3 | 242.6 | -6.5 | B5Ia | 2.45 | -7.51 | 1.02 |
| Alpha Cephei | Alderamin | 21 19 | +62.6 | 101.0 | +9.1 | A7IV | 2.45 | 1.58 | 66.84 |
| Kappa Velorum | Markeb | 09 22 | -55.0 | 275.9 | -3.5 | B2IV | 2.47 | -3.62 | 6.05 |
| Epsilon Cygni | Gienah | 20 46 | +34.0 | 76.0 | -5.7 | K0III | 2.48 | 0.76 | 45.26 |
| Alpha Pegasi | Markab | 23 05 | +15.2 | 88.4 | -40.4 | B9IV | 2.49 | -0.67 | 23.36 |
| Alpha Ceti | Menkar | 03 02 | +4.1 | 173.3 | -45.6 | M2III | 2.54 | -1.61 | 14.82 |
| Zeta Ophiuchi | Han | 16 37 | -10.6 | 6.2 | +23.6 | O9.5V | 2.54 | -3.20 | 7.12 |
| Zeta Centauri | Al Nair al Kent. | 13 56 | -47.3 | 314.2 | +14.2 | B2.5IV | 2.55 | -2.81 | 8.48 |
| Delta Leonis | Zosma | 11 14 | +20.5 | 224.3 | +66.8 | A4V | 2.56 | 1.32 | 56.52 |
| Beta Scorpii | Graffias | 16 05 | -19.8 | 353.1 | +23.7 | B1V+B2V | 2.56 | -3.50 | 6.15 |
| Alpha Leporis | Arneb | 05 33 | -17.8 | 221.0 | -25.1 | F0Ib | 2.58 | -5.40 | 2.54 |
| Delta Centauri | | 12 08 | -50.7 | 295.9 | +11.6 | B2IV | 2.58v | -2.84v | 8.25 |
| Gamma Corvi | Gienah Ghurab | 12 16 | -17.5 | 291.1 | +44.6 | B8III | 2.58 | -0.94 | 19.78 |
| Zeta Sagittarii | Ascella | 19 03 | -29.9 | 6.9 | -15.5 | A2IV+A4V | 2.60 | 0.42 | 36.61 |
| Beta Librae | Zubeneschamali | 15 17 | -9.4 | 352.0 | +39.2 | B8V | 2.61 | -0.84 | 20.38 |
| Alpha Serpentis | Unukalhai | 15 44 | +6.4 | 14.1 | +44.1 | K2III | 2.63 | 0.87 | 44.54 |
| Beta Arietis | Sheratan | 01 55 | +20.8 | 142.4 | -39.7 | A5V | 2.64 | 1.33 | 54.74 |
| Alpha Librae | Zubenelgenubi | 14 51 | -16.0 | 340.4 | +38.0 | A3IV+F4IV | 2.64 | 0.77 | 42.25 |
| Alpha Columbae | Phact | 05 40 | -34.1 | 238.9 | -28.8 | B7IV | 2.65 | -1.93 | 12.16 |
| Theta Aurigae | | 06 00 | +37.2 | 174.4 | +6.8 | A0III+G2V | 2.65 | -0.98 | 18.83 |
| Beta Corvi | Kraz | 12 34 | -23.4 | 297.8 | +39.3 | G5III | 2.65 | -0.51 | 23.34 |
| Delta Cassiopeiae | Ruchbah | 01 26 | +60.2 | 127.2 | -2.4 | A5III | 2.66 | 0.24 | 32.81 |
| Eta Boötis | Muphrid | 13 55 | +18.4 | 5.5 | +73.0 | G0IV | 2.68 | 2.41 | 88.17 |
| Beta Lupi | Ke Kouan | 14 59 | -43.1 | 326.4 | +13.9 | B2III | 2.68 | -3.35 | 6.23 |
| Iota Aurigae | Hassaleh | 04 57 | +33.2 | 170.6 | -6.1 | K3II | 2.69 | -3.29 | 6.37 |
| Mu Velorum | | 10 47 | -49.4 | 283.1 | +8.6 | G5III+G2V | 2.69 | -0.06 | 28.18 |
| Alpha Muscae | | 12 37 | -69.1 | 301.6 | -6.3 | B2V | 2.69 | -2.17 | 10.67 |
| Upsilon Scorpii | Lesath | 17 31 | -37.3 | 351.3 | -1.9 | B2IV | 2.70 | -3.31 | 6.29 |
| Pi Puppis | | 07 17 | -37.1 | 249.0 | -11.3 | K4Ib | 2.71 | -4.92 | 2.98 |

| | | | | | | | | | |
|---|---|---|---|---|---|---|---|---|---|
| Delta Sagittarii | Kaus Meridionalis | 18 21 | -29.8 | 3.0 | -7.2 | K2II | 2.72 | -2.14 | 10.67 |
| Gamma Aquilae | Tarazed | 19 46 | +10.6 | 48.7 | -7.0 | K3II | 2.72 | -3.03 | 7.08 |
| Delta Ophiuchi | Yed Prior | 16 14 | -3.7 | 8.8 | +32.3 | M1III | 2.73 | -0.86 | 19.16 |
| Eta Draconis | Aldhibain | 16 24 | +61.5 | 92.6 | +40.9 | G8III | 2.73 | 0.58 | 37.18 |
| Theta Carinae | | 10 43 | -64.4 | 289.6 | -4.9 | B0V | 2.74 | -2.91 | 7.43 |
| Gamma Virginis | Porrima | 12 42 | -1.5 | 298.1 | +61.3 | F0V+F0V | 2.74 | 2.38 | 84.53 |
| Iota Orionis | Hatysa | 05 35 | -5.9 | 209.5 | -19.7 | O9III | 2.75 | -5.30 | 2.46 |
| Iota Centauri | | 13 21 | -36.7 | 309.5 | +25.8 | A2V | 2.75 | 1.48 | 55.64 |
| Beta Ophiuchi | Cebalrai | 17 43 | +4.6 | 29.2 | +17.3 | K2III | 2.76 | 0.76 | 39.78 |
| Beta Eridani | Kursa | 05 08 | -5.1 | 205.4 | -25.3 | A3III | 2.78 | 0.60 | 36.71 |
| Beta Herculis | Kornephoros | 16 30 | +21.5 | 39.0 | +40.3 | G7III | 2.78 | -0.50 | 22.07 |
| Delta Crucis | | 12 15 | -58.7 | 298.2 | +3.8 | B2IV | 2.79 | -2.45 | 8.96 |
| Beta Draconis | Rastaban | 17 30 | +52.3 | 79.6 | +33.4 | G2II | 2.79 | -2.43 | 9.02 |
| Alpha Canum Venaticorum | Cor Caroli | 12 56 | +38.3 | 118.3 | +78.8 | A0IV+F0V | 2.80v | 0.16v | 29.60 |
| Gamma Lupi | | 15 35 | -41.2 | 333.2 | +11.9 | B2IV-V+B2IV-V | 2.80 | -3.40 | 5.75 |
| Beta Leporis | Nihal | 05 28 | -20.8 | 223.6 | -27.3 | G5III | 2.81 | -0.63 | 20.49 |
| Zeta Herculis | Rutilicus | 16 41 | +31.6 | 52.6 | +40.3 | F9IV+G7V | 2.81 | 2.64 | 92.63 |
| Beta Hydri | | 00 26 | -77.3 | 304.7 | -39.7 | G2IV | 2.82 | 3.45 | 133.78 |
| Tau Scorpii | | 16 36 | -28.2 | 351.6 | +12.8 | B0V | 2.82 | -2.78 | 7.59 |
| Lambda Sagittarii | Kaus Borealis | 18 28 | -25.4 | 7.7 | -6.5 | K1III | 2.82 | 0.95 | 42.20 |
| Gamma Pegasi | Algenib | 00 13 | +15.2 | 109.4 | -46.7 | B2IV | 2.83 | -2.22 | 9.79 |
| Rho Puppis | Turais | 08 08 | -24.3 | 243.2 | +4.5 | F6III | 2.83 | 1.41 | 51.99 |
| Beta Trianguli Australis | | 15 55 | -63.4 | 321.9 | -7.5 | F2IV | 2.83 | 2.38 | 81.24 |
| Zeta Persei | | 03 54 | +31.9 | 162.3 | -16.7 | B1II+B8IV+A2V | 2.84 | -4.55 | 3.32 |
| Beta Arae | | 17 25 | -55.5 | 335.4 | -11.0 | K3Ib-II | 2.84 | -3.49 | 5.41 |
| Alpha Arae | Choo | 17 32 | -49.9 | 340.8 | -8.9 | B2V | 2.84 | -1.51 | 13.46 |
| Eta Tauri | Alcyone | 03 47 | +24.1 | 166.6 | -23.5 | B7III | 2.85 | -2.41 | 8.87 |
| Epsilon Virginis | Vindemiatrix | 13 02 | +11.0 | 312.3 | +73.7 | G8III | 2.85 | 0.37 | 31.90 |
| Delta Capricorni | Deneb Algedi | 21 47 | -16.1 | 37.6 | -46.0 | A5V | 2.85e | 2.49 | 84.58 |
| Alpha Hydri | Head of Hydrus | 01 59 | -61.6 | 289.4 | -53.7 | F0III | 2.86 | 1.16 | 45.74 |
| Delta Cygni | | 19 45 | +45.1 | 78.7 | +10.2 | B9.5III+F1V | 2.86 | -0.74 | 19.07 |
| Mu Geminorum | Tejat | 06 23 | +22.5 | 189.8 | +4.2 | M3III | 2.87v | -1.39v | 14.07 |
| Gamma Trianguli Australis | | 15 19 | -68.7 | 316.5 | -8.4 | A1III | 2.87 | -0.87 | 17.85 |
| Alpha Tucanae | | 22 19 | -60.3 | 330.1 | -48.0 | K3III | 2.87 | -1.05 | 16.42 |
| Theta Eridani | Acamar | 02 58 | -40.3 | 247.9 | -60.7 | A4III+A1V | 2.88 | -0.59 | 20.22 |
| Pi Sagittarii | Albaldah | 19 10 | -21.0 | 15.9 | -13.3 | F2II | 2.88 | -2.77 | 7.41 |
| Beta Canis Minoris | Gomeisa | 07 27 | +08.3 | 209.5 | +11.7 | B8V | 2.89 | -0.70 | 19.16 |
| Pi Scorpii | | 15 59 | -26.1 | 347.2 | +20.2 | B1V+B2V | 2.89 | -2.85 | 7.10 |
| Epsilon Persei | | 03 58 | +40.0 | 157.4 | -10.1 | B0.5V+A2V | 2.90 | -3.19 | 6.06 |
| Sigma Scorpii | Alniyat | 16 21 | -25.6 | 351.3 | +17.0 | B1III | 2.90v | -3.86v | 4.44 |
| Beta Cygni | Albireo | 19 31 | +28.0 | 62.1 | +4.6 | K3II+B8V+B9V | 2.90 | -2.31 | 8.46 |
| Beta Aquarii | Sadalsuud | 21 32 | -05.6 | 48.0 | -37.9 | G0Ib | 2.90 | -3.47 | 5.33 |
| Gamma Persei | | 03 05 | +53.5 | 142.1 | -4.3 | G8III+A2V | 2.91 | -1.57 | 12.72 |
| Upsilon Carinae | | 09 47 | -65.1 | 285.0 | -8.8 | A7Ib+B7III | 2.92 | -5.56 | 2.01 |
| Eta Pegasi | Matar | 22 43 | +30.2 | 92.5 | -25.0 | G2II-III+F0V | 2.93 | -1.16 | 15.18 |
| Tau Puppis | | 06 50 | -50.6 | 260.2 | -20.9 | K1III | 2.94 | -0.80 | 17.85 |
| Delta Corvi | Algorel | 12 30 | -16.5 | 295.5 | +46.0 | B9.5V | 2.94 | 0.79 | 37.11 |
| Alpha Aquarii | Sadalmelik | 22 06 | -00.3 | 59.9 | -42.1 | G2Ib | 2.95 | -3.88 | 4.30 |
| Gamma Eridani | Zaurak | 03 58 | -13.5 | 205.2 | -44.5 | M1III | 2.97 | -1.19 | 14.75 |
| Zeta Tauri | Alheka | 05 38 | +21.1 | 185.7 | -5.6 | B4III | 2.97 | -2.56 | 7.82 |
| Epsilon Leonis | Ras Elased Austr. | 09 46 | +23.8 | 206.8 | +48.2 | G1II | 2.97 | -1.46 | 13.01 |
| Gamma$^2$ Sagittarii | Alnasl | 18 06 | -30.4 | 0.9 | -4.5 | K0III | 2.98 | 0.63 | 33.94 |
| Gamma Hydrae | | 13 19 | -23.2 | 311.1 | +39.3 | G8III | 2.99 | -0.05 | 24.69 |
| Iota$^1$ Scorpii | | 17 48 | -40.1 | 350.6 | -6.1 | F2Ia | 2.99 | -5.71 | 1.82 |
| Zeta Aquilae | Deneb el Okab | 19 05 | +13.9 | 46.9 | +3.3 | A0V | 2.99 | 0.96 | 39.18 |
| Beta Trianguli | | 02 10 | +35.0 | 140.6 | -25.2 | A5III | 3.00 | 0.09 | 26.24 |
| Psi Ursae Majoris | | 11 10 | +44.5 | 165.8 | +63.2 | K1III | 3.00 | -0.27 | 22.21 |
| Gamma Ursae Minoris | Pherkad Major | 15 21 | +71.8 | 108.5 | +40.8 | A3II | 3.00 | -2.84 | 6.79 |
| Mu$^1$ Scorpii | | 16 52 | -38.0 | 346.1 | +3.9 | B1.5V+B6.5V | 3.00 | -4.01 | 3.97 |
| Gamma Gruis | | 21 54 | -37.4 | 6.1 | -51.5 | B8III | 3.00 | -0.97 | 16.07 |

| | | | | | | | | | |
|---|---|---|---|---|---|---|---|---|---|
| Delta Persei | | 03 43 | +47.8 | 150.3 | -5.8 | B5III | 3.01 | -3.04 | 6.18 |
| Zeta Canis Majoris | Phurad | 06 20 | -30.1 | 237.5 | -19.4 | B2.5V | 3.02 | -2.05 | 9.70 |
| Omicron$^2$ Canis Majoris | | 07 03 | -23.8 | 235.6 | -8.2 | B3Ia | 3.02 | -6.46 | 1.27 |
| Epsilon Corvi | Minkar | 12 10 | -22.6 | 290.6 | +39.3 | K2II | 3.02 | -1.82 | 10.75 |
| Epsilon Aurigae | Almaaz | 05 02 | +43.8 | 162.8 | +1.2 | F0Ia | 3.03e | -5.95 | 1.60 |
| Beta Muscae | | 12 46 | -68.1 | 302.5 | -5.2 | B2V+B3V | 3.04 | -1.86 | 10.48 |
| Gamma Boötis | Seginus | 14 32 | +38.3 | 67.3 | +66.2 | A7III | 3.04 | 0.96 | 38.29 |
| Beta Capricorni | Dabih | 20 21 | -14.8 | 29.2 | -26.4 | G5II+A0V | 3.05 | -2.07 | 9.48 |
| Epsilon Geminorum | Mebsuta | 06 44 | +25.1 | 189.5 | +9.6 | G8Ib | 3.06 | -4.15 | 3.61 |
| Mu Ursae Majoris | Tania Australis | 10 22 | +41.5 | 177.9 | +56.4 | M0III | 3.06e | -1.35 | 13.11 |
| Delta Draconis | Tais | 19 13 | +67.7 | 98.7 | +23.0 | G9III | 3.07 | 0.63 | 32.54 |
| Eta Sagittarii | | 18 18 | -36.8 | 356.4 | -9.7 | M3.5III | 3.10 | -0.20 | 21.87 |
| Zeta Hydrae | | 08 55 | +05.9 | 222.3 | +30.2 | G9III | 3.11 | -0.21 | 21.64 |
| Nu Hydrae | | 10 50 | -16.2 | 265.1 | +37.6 | K2III | 3.11 | -0.03 | 23.54 |
| Lambda Centauri | | 11 36 | -63.0 | 294.5 | -1.4 | B9III | 3.11 | -2.39 | 7.96 |
| Alpha Indi | Persian | 20 38 | -47.3 | 352.6 | -37.2 | K0III | 3.11 | 0.65 | 32.21 |
| Beta Columbae | Wazn | 05 51 | -35.8 | 241.3 | -27.1 | K2III | 3.12 | 1.02 | 37.94 |
| Iota Ursae Majoris | Talita | 08 59 | +48.0 | 171.5 | +40.8 | A7IV | 3.12 | 2.29 | 68.32 |
| Zeta Arae | | 16 59 | -56.0 | 332.8 | -8.2 | K3II | 3.12 | -3.11 | 5.68 |
| Delta Herculis | Sarin | 17 15 | +24.8 | 46.8 | +31.4 | A3IV | 3.12 | 1.21 | 41.55 |
| Kappa Centauri | Ke Kwan | 14 59 | -42.1 | 326.9 | +14.8 | B2IV | 3.13 | -2.96 | 6.05 |
| Alpha Lyncis | | 09 21 | +34.4 | 190.2 | +44.7 | K7III | 3.14 | -1.02 | 14.69 |
| N Velorum | | 09 31 | -57.0 | 278.2 | -4.1 | K5III | 3.16 | -1.15 | 13.72 |
| Pi Herculis | | 17 15 | +36.8 | 60.7 | +34.3 | K3II | 3.16 | -2.10 | 8.89 |
| Nu Puppis | | 06 38 | -43.2 | 251.9 | -20.5 | B8III | 3.17 | -2.39 | 7.71 |
| Theta Ursae Majoris | Al Haud | 09 33 | +51.7 | 165.5 | +45.7 | F6IV | 3.17 | 2.52 | 74.15 |
| Zeta Draconis | Aldhibah | 17 09 | +65.7 | 96.0 | +35.0 | B6III | 3.17 | -1.92 | 9.60 |
| Phi Sagittarii | | 18 46 | -27.0 | 8.0 | -10.8 | B8III | 3.17 | -1.08 | 14.14 |
| Eta Aurigae | Hoedus II | 05 07 | +41.2 | 165.4 | +0.3 | B3V | 3.18 | -0.96 | 14.87 |
| Alpha Circini | | 14 43 | -65.0 | 314.3 | -4.6 | F0V+K5V | 3.18 | 2.11 | 60.97 |
| Pi$^3$ Orionis | Tabit | 04 50 | +07.0 | 191.5 | -23.1 | F6V | 3.19 | 3.67 | 124.60 |
| Epsilon Leporis | | 05 05 | -22.4 | 223.3 | -32.7 | K5III | 3.19 | -1.02 | 14.39 |
| Kappa Ophiuchi | | 16 58 | +09.4 | 28.4 | +29.5 | K2III | 3.19 | 1.09 | 37.99 |
| G Scorpii | | 17 50 | -37.0 | 353.5 | -4.9 | K2III | 3.19 | 0.24 | 25.71 |
| Zeta Cygni | | 21 13 | +30.2 | 76.8 | -12.5 | G8III | 3.21 | -0.12 | 21.62 |
| Gamma Cephei | Errai | 23 39 | +77.6 | 119.0 | +15.3 | K1IV | 3.21 | 2.51 | 72.50 |
| Delta Lupi | | 15 21 | -40.6 | 331.3 | +13.8 | B1.5IV | 3.22 | -2.75 | 6.39 |
| Epsilon Ophiuchi | Yed Posterior | 16 18 | -04.7 | 8.6 | +30.8 | G9III | 3.23 | 0.64 | 30.34 |
| Eta Serpentis | Alava | 18 21 | -02.9 | 26.9 | +5.4 | K0III-IV | 3.23 | 1.84 | 52.81 |
| Beta Cephei | Alphirk | 21 29 | +70.6 | 107.5 | +14.0 | B2III | 3.23v | -3.08v | 5.48 |
| Alpha Pictoris | | 06 48 | -61.9 | 271.9 | -24.1 | A7III | 3.24 | 0.83 | 32.96 |
| Theta Aquilae | | 20 11 | -00.8 | 41.6 | -18.1 | B9.5III | 3.24 | -1.48 | 11.36 |
| Sigma Puppis | | 07 29 | -43.3 | 255.7 | -11.9 | K5III+G5V | 3.25 | -0.51 | 17.74 |
| Pi Hydrae | | 14 06 | -26.7 | 323.0 | +33.3 | K2III | 3.25 | 0.79 | 32.17 |
| Sigma Librae | Brachium | 15 04 | -25.3 | 337.2 | +28.6 | M3III | 3.25 | -1.51 | 11.17 |
| Gamma Lyrae | Sulaphat | 18 59 | +32.7 | 63.3 | +12.8 | B9II | 3.25 | -3.20 | 5.14 |
| Gamma Hydri | | 03 47 | -74.2 | 289.1 | -37.8 | M2III | 3.26 | -0.83 | 15.23 |
| Delta Andromedae | | 00 39 | +30.9 | 119.9 | -31.9 | K3III | 3.27 | 0.81 | 32.19 |
| Theta Ophiuchi | | 17 22 | -25.0 | 0.5 | +6.6 | B2IV | 3.27 | -2.92 | 5.79 |
| Delta Aquarii | Skat | 22 55 | -15.8 | 49.6 | -60.7 | A3III | 3.27 | -0.18 | 20.44 |
| Mu Leporis | | 05 13 | -16.2 | 217.3 | -28.9 | B9IV | 3.29v | -0.47v | 17.69 |
| Omega Carinae | | 10 14 | -70.0 | 290.2 | -11.2 | B8III | 3.29 | -1.99 | 8.81 |
| Iota Draconis | Edasich | 15 25 | +59.0 | 94.0 | +48.6 | K2III | 3.29 | 0.81 | 31.92 |
| Alpha Doradus | | 04 34 | -55.0 | 263.8 | -41.4 | A0IV+B9IV | 3.30 | -0.36 | 18.56 |
| p Carinae | | 10 32 | -61.7 | 287.2 | -3.2 | B4V | 3.30 | -2.62 | 6.56 |
| Mu Centauri | | 13 50 | -42.5 | 314.2 | +19.1 | B2IV-V | 3.30v | -2.74v | 6.19 |
| Eta Geminorum | Propus | 06 15 | +22.5 | 188.9 | +2.5 | M3III | 3.31v | -1.84v | 9.34 |
| Alpha Herculis | Rasalgethi | 17 15 | +14.4 | 35.5 | +27.8 | M5III+G5III | 3.31v | -2.04v | 8.53 |
| Gamma Arae | | 17 25 | -56.4 | 334.6 | -11.5 | B1III | 3.31 | -4.40 | 2.87 |
| Beta Phoenicis | | 01 06 | -46.7 | 295.5 | -70.2 | G8III | 3.32 | -0.55 | 16.9 |

| | | | | | | | | | |
|---|---|---|---|---|---|---|---|---|---|
| Rho Persei | Gorgonea Tertia | 03 05 | +38.8 | 149.6 | -17.0 | M3III | 3.32v | -1.67v | 10.03 |
| Delta Ursae Majoris | Megrez | 12 15 | +57.0 | 132.6 | +59.4 | A3V | 3.32 | 1.33 | 40.05 |
| Eta Scorpii | | 17 12 | -43.2 | 344.4 | -2.3 | F3III-IV | 3.32 | 1.61 | 45.56 |
| Nu Ophiuchi | | 17 59 | -09.8 | 18.2 | +7.0 | K0III | 3.32 | -0.03 | 21.35 |
| Tau Sagittarii | | 19 07 | -27.7 | 9.3 | -15.4 | K1III | 3.32 | 0.48 | 27.09 |
| Alpha Reticuli | | 04 14 | -62.5 | 274.3 | -41.7 | G8III | 3.33 | -0.17 | 19.98 |
| Theta Leonis | Chort | 11 14 | +15.4 | 235.4 | +64.6 | A2III | 3.33 | -0.35 | 18.36 |
| Xi Puppis | Asmidiske | 07 49 | -24.9 | 241.5 | +0.6 | G5Ib | 3.34 | -4.74 | 2.42 |
| Epsilon Cassiopeiae | Segin | 01 54 | +63.7 | 129.9 | +1.7 | B2III | 3.35 | -2.31 | 7.38 |
| Eta Orionis | Algjebbah | 05 24 | -02.4 | 204.9 | -20.4 | B1V+B2V | 3.35 | -3.86 | 3.62 |
| Xi Geminorum | Alzirr | 06 45 | +12.9 | 200.7 | +4.5 | F5IV | 3.35 | 2.13 | 57.02 |
| Omicron Ursae Majoris | Muscida | 08 30 | +60.7 | 156.0 | +35.4 | G5III | 3.35v | -0.40v | 17.76 |
| Delta Aquilae | | 19 25 | +03.1 | 39.6 | -6.1 | F2IV | 3.36 | 2.43 | 65.05 |
| Epsilon Lupi | | 15 23 | -44.7 | 329.2 | +10.3 | B2IV-V | 3.37 | -2.58 | 6.47 |
| Zeta Virginis | Heze | 13 35 | -00.6 | 325.3 | +60.4 | A3V | 3.38 | 1.62 | 44.55 |
| Epsilon Hydrae | | 08 47 | +06.4 | 220.7 | +28.5 | G5III+A8V+F7V | 3.38 | 0.29 | 24.13 |
| Lambda Orionis | Meissa | 05 35 | +09.9 | 195.1 | -12.0 | O8III | 3.39 | -4.16 | 3.09 |
| q Carinae | | 10 17 | -61.3 | 285.5 | -3.8 | K3II | 3.39 | -3.38 | 4.43 |
| Delta Virginis | Auva | 12 56 | +03.4 | 305.5 | +66.3 | M3III | 3.39 | -0.57 | 16.11 |
| Zeta Cephei | | 22 11 | +58.2 | 103.1 | +1.7 | K1II | 3.39 | -3.35 | 4.49 |
| Theta² Tauri | | 04 29 | +15.9 | 180.4 | -22.0 | A7III | 3.40 | 0.10 | 21.89 |
| Gamma Phoenicis | | 01 28 | -43.3 | 280.5 | -72.2 | K5III | 3.41 | -0.87 | 13.94 |
| Lambda Tauri | | 04 01 | +12.5 | 178.4 | -29.4 | B3V+A4IV | 3.41e | -1.87 | 8.81 |
| Nu Centauri | | 13 50 | -41.7 | 314.4 | +19.9 | B2IV | 3.41 | -2.41 | 6.87 |
| Zeta Lupi | | 15 12 | -52.1 | 323.8 | +5.0 | G8III | 3.41 | 0.65 | 28.06 |
| Eta Cephei | | 20 45 | +61.8 | 097.9 | +11.6 | K0IV | 3.41 | 2.63 | 69.73 |
| Zeta Pegasi | Homam | 22 41 | +10.8 | 078.9 | -40.7 | B8.5V | 3.41 | -0.62 | 15.64 |
| Alpha Trianguli | Mothallah | 01 53 | +29.6 | 138.6 | -31.4 | F6IV | 3.42 | 1.95 | 50.87 |
| Eta Lupi | | 16 00 | -38.4 | 338.8 | +11.0 | B2.5IV+A5V | 3.42 | -2.48 | 6.61 |
| Mu Herculis | | 17 46 | +27.7 | 052.4 | +25.6 | G5IV | 3.42 | 3.80 | 119.05 |
| Beta Pavonis | | 20 45 | -66.2 | 329.0 | -36.0 | A7III | 3.42 | 0.29 | 23.71 |
| a Carinae | | 09 11 | -58.9 | 277.7 | -7.4 | B2IV | 3.43 | -2.11 | 7.79 |
| Zeta Leonis | Adhafera | 10 17 | +23.4 | 210.2 | +55.0 | F0II-III | 3.43 | -1.08 | 12.56 |
| Lambda Aquilae | Althalimain | 19 06 | -04.9 | 030.3 | -5.5 | B9V | 3.43 | 0.51 | 26.05 |
| Lambda Ursae Majoris | Tania Borealis | 10 17 | +42.9 | 175.9 | +55.1 | A2IV | 3.45 | 0.38 | 24.27 |
| Beta Lyrae | Sheliak | 18 50 | +33.4 | 63.2 | +14.8 | B8II | 3.45e | -3.71 | 3.70 |
| Eta Cassiopeiae | Achird | 00 49 | +57.8 | 122.6 | -5.1 | G0V+K7V | 3.46 | 4.59 | 167.99 |
| Eta Ceti | Dheneb | 01 09 | -10.2 | 137.2 | -72.6 | K2III | 3.46 | 0.67 | 27.73 |
| Chi Carinae | | 07 57 | -53.0 | 266.7 | -12.3 | B3IV | 3.46 | -1.91 | 8.43 |
| Delta Bootis | | 15 16 | +33.3 | 053.1 | +58.4 | G8III | 3.46 | 0.69 | 27.94 |
| Gamma Ceti | Kaffaljidhma | 02 43 | +03.2 | 168.9 | -49.4 | A3V+F3V+K5V | 3.47 | 1.47 | 39.78 |
| Eta Leonis | | 10 07 | +16.8 | 219.5 | +50.8 | A0Ib | 3.48 | -5.60 | 1.53 |
| Eta Herculis | | 16 43 | +38.9 | 062.3 | +40.9 | G8III | 3.48 | 0.80 | 29.11 |
| Tau Ceti | | 01 44 | -15.9 | 173.1 | -73.4 | G8V | 3.49 | 5.68 | 274.17 |
| Sigma Canis Majoris | | 07 02 | -27.9 | 239.2 | -10.3 | K7Ib | 3.49 | -4.37 | 2.68 |
| Nu Ursae Majoris | Alula Borealis | 11 18 | +33.1 | 190.7 | +69.1 | K3II | 3.49 | -2.07 | 7.74 |
| Beta Bootis | Nekkar | 15 02 | +40.4 | 067.6 | +60.0 | G8III | 3.49 | -0.64 | 14.91 |
| Alpha Telescopii | | 18 27 | -46.0 | 348.7 | -15.2 | B3IV | 3.49 | -0.93 | 13.08 |
| Epsilon Gruis | | 22 49 | -51.3 | 338.3 | -56.5 | A3V | 3.49 | 0.49 | 25.16 |
| Kappa Canis Majoris | | 06 50 | -32.5 | 242.4 | -14.5 | B1.5IV | 3.50 | -3.42 | 4.13 |
| Delta Geminorum | Wasat | 07 20 | +22.0 | 196.0 | +15.9 | F2IV+K3V | 3.50 | 2.22 | 55.45 |
| Iota Cephei | | 22 50 | +66.2 | 111.1 | +6.2 | K0III | 3.50 | 0.76 | 28.27 |
| Gamma Sagittae | | 19 59 | +19.5 | 58.0 | -5.2 | K5III | 3.51 | -1.11 | 11.90 |
| Mu Pegasi | Sadalbari | 22 50 | +24.6 | 90.7 | -30.6 | G8III | 3.51 | 0.74 | 27.95 |
| Delta Eridani | Rana | 03 43 | -09.8 | 198.1 | -46.0 | K0IV | 3.52 | 3.74 | 110.58 |
| Omicron Leonis | Subra | 09 41 | +09.9 | 224.6 | +42.1 | A9V+F6V | 3.52 | 0.43 | 24.12 |
| Phi Velorum | Tseen Ke | 09 57 | -54.6 | 279.4 | +0.1 | B5Ib | 3.52 | -5.34 | 1.69 |
| Xi² Sagittarii | | 18 58 | -21.1 | 14.6 | -10.8 | K0II | 3.52 | -1.77 | 8.76 |
| Theta Pegasi | Baham | 22 10 | +06.2 | 67.4 | -38.7 | A2V | 3.52 | 1.16 | 33.77 |
| Epsilon Tauri | Ain | 04 29 | +19.2 | 177.6 | -19.9 | K0III | 3.53 | 0.15 | 21.04 |

| | | | | | | | | | |
|---|---|---|---|---|---|---|---|---|---|
| Beta Cancri | Tarf | 08 17 | +09.2 | 214.3 | +23.1 | K4III | 3.53 | -1.24 | 11.23 |
| Xi Hydrae | | 11 33 | -31.9 | 284.1 | +28.1 | G8III | 3.54 | 0.55 | 25.23 |
| Mu Serpentis | | 15 50 | -03.4 | 4.6 | +37.3 | A0V | 3.54 | 0.14 | 20.94 |
| Xi Serpentis | | 17 38 | -15.4 | 10.6 | +8.7 | F0III | 3.54 | 0.99 | 30.93 |

Column 1: Catalog name for the star.

Column 2: Common name of the star.

Column 3: Right Ascension in hours and minutes for epoch 2000.

Column 4: Declination in degrees for epoch 2000.

Column 5: Galactic longitude of the star.

Column 6: Galactic latitude of the star.

Column 7: Spectral classification of the main stars in the system.

Column 8: Apparent visual magnitude of the star. A letter 'v' means the magnitude varies by more than 0.1. A letter 'e' means that the star is an eclipsing binary.

Column 9: Absolute magnitude of the star. A letter 'v' means the magnitude varies by more than 0.1.

Column 10: The Hipparcos parallax of the star (milli-arcseconds).

References:

ESA, (1997), The Hipparcos Catalogue.

Hoffleit D, Warren Jr W, (1991), Harvard Revised Bright Star Catalogue, 5th Edition.

## THE ANGULAR AND TIME MEASURE

One can measure angles on the sky either by means of degrees or angular hours, minutes and seconds. The globe has been divided into 24h (time hours) and therefore 24h is $360^{\circ}$ (arc degrees), 1h is $60'$ (arc minutes) and 1 minute (time minute) is $60''$ (arc seconds). These divisions also pertain to the celestial sphere.

Table and converter

| time | arc |
|---|---|
| **1h** | **$15^{\circ}$** |
| **1m** | **15'** |

| arc | time |
|---|---|
| **$1^{\circ}$** | **4m** |
| **1'** | **4s** |

You can also apply the following relations:

$$1^{\circ} \leftrightarrow \left(\frac{1}{15}\right)^{h} = 4^{m} = 240^{s}$$

$$1^{h} \leftrightarrow 15^{\circ} = 900' = 54000''$$

Exercises:

1. Change the time measure into angular (arcwise) one

| Time measure | Arc (angular) measure |
|---|---|
| **0h 39m** | |
| **1h 10m** | |
| **2h 50m** | |
| **4h 15m** | |
| **5h 13m** | |
| **7h 53m** | |

2. Change the angular (arcwise) measure into the time one

| Arc (angular) measure | Time measure |
|---|---|
| **12° 5'** | |
| **30° 10'** | |
| **53° 48'** | |
| **66° 55'** | |
| **100° 10'** | |
| **165° 48'** | |

3. The latitude of Phoenix College is $33^{\circ}\, 30'\, 29''\, N$. Convert it into time measure. Use the space below in order to show your calculations.

---

4. The longitude of Phoenix College is $-112^{\circ}\ 5'\ 11''\ W$. Express it into time hours, minutes and seconds. Use the space below for calculations.

## ADDITION AND SUBTRACTION OF ANGLES

In order to add or subtract angles expressed in angular measure (i.e. in the form of A° B′ C″), one should add/subtract degrees, minutes and seconds separately. In the event that the number of minutes or seconds is greater than 60, one should convert it (the number) into larger units (leaving the remainder in the smaller unit). In the event that it is lower than 0, one must change one larger unit into 60 smaller ones and sum them up.

E.g.:

- 58°25'+13°40'=71°+65'=72°5'
- 58°25'-13°40'=45°+(-15')=44°45'

Exercises:

1. Add angles:

a) 17°21'+0°40'

b) 18°33'+90°34'

c) 128°54'+12°6'

## CONSTELLATIONS

The starry sky has been divided into 88 constellations. Each of them has its own name (e.g. Cassiopeia, Ursa Major, Corona Australis, etc.), as well as its own abbreviation. Bright or interesting stars have their own names; for example, Sirius, Vega, and Betelgeuse. However, the brightest stars in constellations are denoted with Greek letters (lower-case). The sequence of alphabet letters correspond with the star brightness.

THE CONSTELLATIONS

| Latin Name | English Name | Abr. | Surface $(°)^2$ | Bright.(mag) | Localization | Hemisphere |
|---|---|---|---|---|---|---|
| Andromeda | Chained Lady | And | 722,278 | 2,06 | 90° N - 37° S | N |
| Aries | Ram | Ari | 441,395 | 2,01 | 90° N - 59° S | N |
| Gemini | Twins | Gem | 513,761 | 1,16 | 90° N - 60° S | N |
| Taurus | Bull | Tau | 797,249 | 0,85 | 90° N - 65° S | N |
| Cepheus | King | Cep | 587,787 | 2,45 | 90° N - 10° S | N |
| Centaurus | Centaur | Cen | 1060,422 | -0,01 | 25° N - 90° S | S |
| Circinus | Compasses | Cir | 93,353 | 3,19 | 20° N - 90° S | S |
| Delphinus | Dolphin | Del | 188,549 | 3,63 | 90° N - 70° S | N |
| Eridanus | River | Eri | 1137,919 | 0,45 | 32° N - 89° S | S |
| Phoenix | Phoenix | Phe | 469,319 | 2,39 | 32° N - 90° S | S |
| Columba | Dove | Col | 270,184 | 2,65 | 41° N - 90° S | S |
| Mensa | Table | Men | 153,484 | 5,09 | 05° N - 90° S | S |
| Hercules | Hercules | Her | 1225,148 | 2,78 | 90° N - 39° S | N |
| Hydra | Water Serpent | Hya | 1302,844 | 1,98 | 55° N - 83° S | S/N |
| Indus | Indian | Ind | 294,006 | 3,11 | 16° N - 90° S | S |
| Lacerta | Lizard | Lac | 200,688 | 3,77 | 90° N - 40° S | N |
| Monoceros | Unicorn | Mon | 481,569 | 3,76 | 75° N - 85° S | N/S |
| Chamaeleon | Chameleon | Cha | 131,592 | 4,05 | 07° N - 90° S | S |
| Cassiopeia | Queen | Cas | 598,407 | 2,24 | 90° N - 12° S | N |
| Carina | Keel | Car | 494,184 | -0,62 | 14° N - 90° S | S |
| Pyxis | Compass | Pyx | 220,833 | 3,68 | 53° N - 90° S | S |
| Corona Borealis | Northern Crown | CrB | 178,71 | 2,22 | 90° N - 50° S | N |
| Corona Australis | Southern Crown | CrA | 127,696 | 4,1 | 44° N - 90° S | S |
| Capricornus | Water Goat | Cap | 413,947 | 2,73 | 62° N - 90° S | S |

| Latin Name | English Name | Abr. | Surface (°)² | Bright.(mag) | Localization | Hemisphere |
|---|---|---|---|---|---|---|
| Corvus | Crow, Raven | Crv | 183,801 | 2,59 | 65° N - 90° S | S |
| Crux | Southern Cross | Cru | 68,447 | 0,77 | 25° N - 90° S | S |
| Cygnus | Swan | Cyg | 803,983 | 1,25 | 90° N - 29° S | N |
| Leo | Lion | Leo | 946,964 | 1,36 | 83° N - 57° S | N |
| Vulpecula | Fox | Vul | 268,165 | 4,44 | 90° N - 55° S | N |
| Lyra | Lyre | Lyr | 286,476 | 0,03 | 90° N - 29° S | N |
| Ursa Minor | Smaller Bear | UMi | 255,864 | 1,97 | 90° N - 10° S | N |
| Pictor | Easel | Pic | 246,739 | 3,27 | 25° N - 90° S | S |
| Leo Minor | Smaller Lion | LMi | 231,956 | 3,83 | 90° N - 48° S | N |
| Canis Minor | Smaller Dog | CMi | 183,367 | 0,4 | 89° N - 77° S | N |
| Hydrus | Water Snake | Hyi | 243,035 | 2,82 | 08° N - 90° S | S/N |
| Microscopium | Microscope | Mic | 209,513 | 4,67 | 45° N - 90° S | S |
| Musca | Fly | Mus | 138,355 | 2,69 | 10° N - 90° S | S |
| Octans | Octant | Oct | 291,045 | 3,76 | 05° N - 90° S | S |
| Ara | Altar | Ara | 237,057 | 2,85 | 25° N - 90° S | S |
| Orion | Hunter; Orion | Ori | 594,12 | 0,12 | 85° N - 75° S | N/S |
| Aquila | Eagle | Aql | 652,473 | 0,77 | 78° N - 71° S | N/S |
| Virgo | Maiden | Vir | 1294,428 | 0,98 | 67° N - 76° S | S/N |
| Pavo | Peacock | Pav | 377,666 | 1,94 | 15° N - 90° S | S |
| Pegasus | Winged Horse | Peg | 1120,794 | 2,39 | 90° N - 65° S | N |
| Perseus | Hero; Perseus | Per | 614,997 | 1,79 | 90° N - 35° S | N |
| Fornax | Furnace | For | 397,502 | 3,87 | 50° N - 90° S | S |
| Antilia | Air Pump | Ant | 238,901 | 4,25 | 50° N - 90° S | S |
| Canes Venatici | Hunting Dogs | CVn | 465,194 | 2,9 | 90° N - 38° S | N |
| Apus | Bird of Paradise | Aps | 206,327 | 3,83 | 07° N - 90° S | S |
| Crater | Cup | Crt | 282,398 | 3,56 | 65° N - 90° S | S |
| Cancer | Crab | Cnc | 505,872 | 3,52 | 90° N - 60° S | N |
| Puppis | Stern | Pup | 673,434 | 2,06 | 39° N - 90° S | S |
| Volans | Flying Fish | Vol | 141,354 | 3,77 | 14° N - 90° S | S |
| Piscis Austrinus | Southern Fish | PsA | 245,375 | 1,16 | 50° N - 90° S | S |
| Pisces | Fishes | Psc | 889,417 | 3,62 | 84° N - 56° S | N/S |
| Caelum | Engraving Tool | Cae | 124,865 | 4,45 | 41° N - 90° S | S |
| Lynx | Lynx | Lyn | 545,386 | 3,13 | 90° N - 35° S | N |

| Latin Name | English Name | Abr. | Surface (°)² | Bright.(mag) | Localization | Hemisphere |
|---|---|---|---|---|---|---|
| Sculptor | Sculptor's Studio | Scl | 474,764 | 4,31 | 50° N - 90° S | S |
| Sextans | Sextans | Sex | 313,515 | 4,49 | 78° N - 83° S | S |
| Reticulum | Net | Ret | 113,936 | 3,33 | 23° N - 90° S | S |
| Scorpius | Scorpion | Sco | 496,783 | 1,06 | 44° N - 90° S | S |
| Draco | Dragon | Dra | 1082,952 | 2,23 | 90° N - 15° S | N |
| Sagitta | Arrow | Sge | 79,923 | 3,51 | 90° N - 70° S | N |
| Sagittarius | Archer | Sgr | 867,432 | 1,79 | 55° N - 90° S | S |
| Scutum | Shield | Sct | 109,114 | 3,85 | 74° N - 64° S | S |
| Telescopium | Telescope | Tel | 251,512 | 3,51 | 33° N - 90° S | S |
| Triangulum | Triangle | Tri | 131,847 | 3 | 90° N - 50° S | N |
| Triangulum Australe | Southern Triangle | TrA | 109,978 | 1,91 | 20° N - 90° S | S |
| Tucana | Toucan | Tuc | 294,557 | 2,86 | 15° N - 90° S | S |
| Libra | Scales | Lib | 538,052 | 2,61 | 60° N - 90° S | S |
| Coma Berenices | Berenices Hair | Com | 386,475 | 4,26 | 90° N - 56° S | N |
| Serpens | Serpent | Ser | 636,928 | 2,63 | 74° N - 64° S | N/S |
| Norma | Square | Nor | 165,29 | 4,02 | 30° N - 90° S | S |
| Ophiuchus | Serpent Bearer | Oph | 948,34 | 2,08 | 80° N - 80° S | N/S |
| Ursa Major | Greater Bear | UMa | 1279,66 | 1,76 | 90° N - 17° S | N |
| Canis Major | Larger Dog | CMa | 380,118 | -1,44 | 57° N - 90° S | S |
| Cetus | Whale/Sea Monster | Cet | 1231,411 | 2,04 | 70° N - 90° S | S/N |
| Lupus | Wolf | Lup | 333,683 | 2,3 | 35° N - 90° S | S |
| Aquarius | Water Bearer | Aqr | 979,854 | 2,9 | 65° N - 87° S | N/S |
| Bootes | Herdsman | Boo | 906,831 | -0,04 | 90° N - 50° S | N |
| Auriga | Charioteer | Aur | 657,438 | 0,08 | 90° N - 34° S | N |
| Vela | Sails | Vel | 499,649 | 1,75 | 33° N - 90° S | S |
| Lepus | Hare | Lep | 290,291 | 2,58 | 60° N - 90° S | S |
| Horologium | Clock | Hor | 248,885 | 3,85 | 20° N - 90° S | S |
| Dorado | Swordfish | Dor | 179,173 | 3,27 | 20° N - 90° S | S |
| Equuleus | Little Horse | Equ | 71,641 | 3,92 | 90° N - 77° S | N |
| Grus | Crane | Gru | 365,513 | 1,74 | 35° N - 90° S | S |
| Camelopardalis | Giraffe | Cam | 756,828 | 4,03 | 90° N - 37° S | N |

## GREEK ALPHABET

| Letter | Upper case | Lower-case | Letter | Upper-case | Lower-case |
|---|---|---|---|---|---|
| alpha | Α | α | nu | Ν | ν |
| beta | Β | β | xi | Ξ | ξ |
| gamma | Γ | γ | omicron | Ο | ο |
| delta | Δ | δ | pi | Π | π |
| epsilon | Ε | ε | rho | Ρ | ρ |
| zeta | Ζ | ζ | sigma | Σ | σ |
| eta | Η | η | tau | Τ | τ |
| theta | Θ | θ | upsilon | Υ | υ |
| iota | Ι | ι | phi | Φ | φ |
| kappa | Κ | κ | chi | Χ | χ |
| lambda | Λ | λ | psi | Ψ | ψ |
| mu | Μ | μ | omega | Ω | ω |

Exercises:

1. On the base of the constellation charts find five constellations. Draft their shapes and point out the brightest stars in each of them.

First constellation name:

Second constellation name:

Third constellation name:

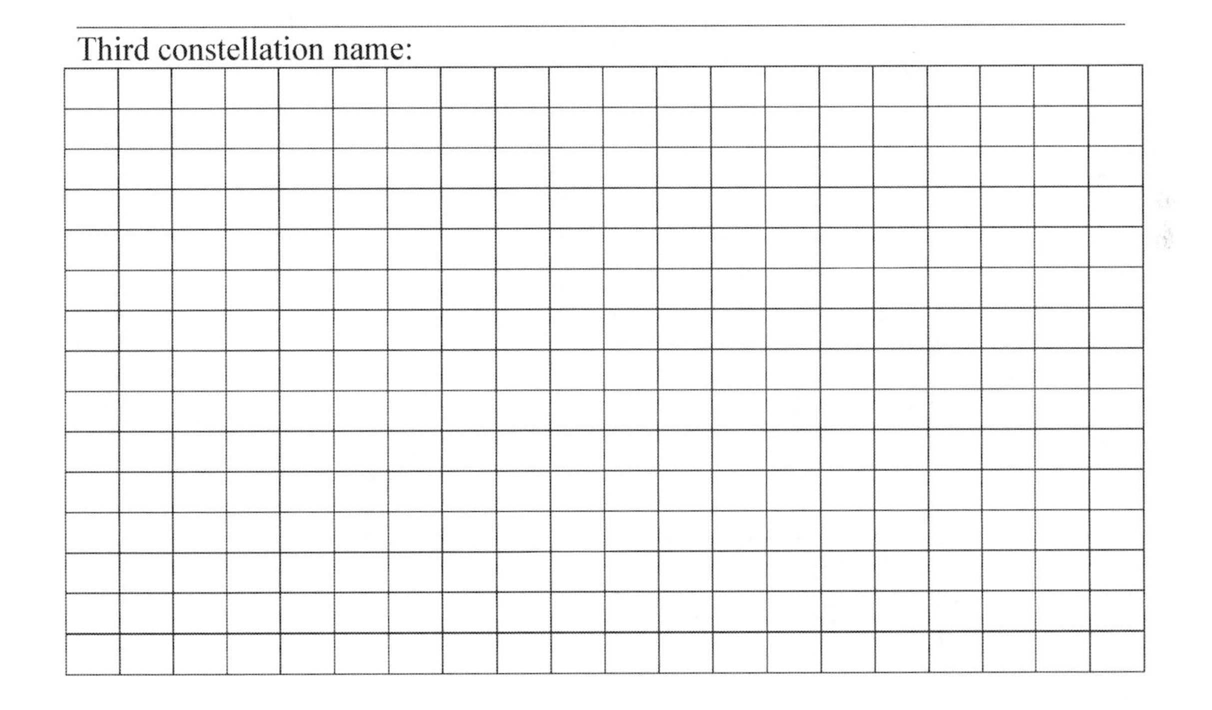

Fourth constellation name:

Fifth constellation name:

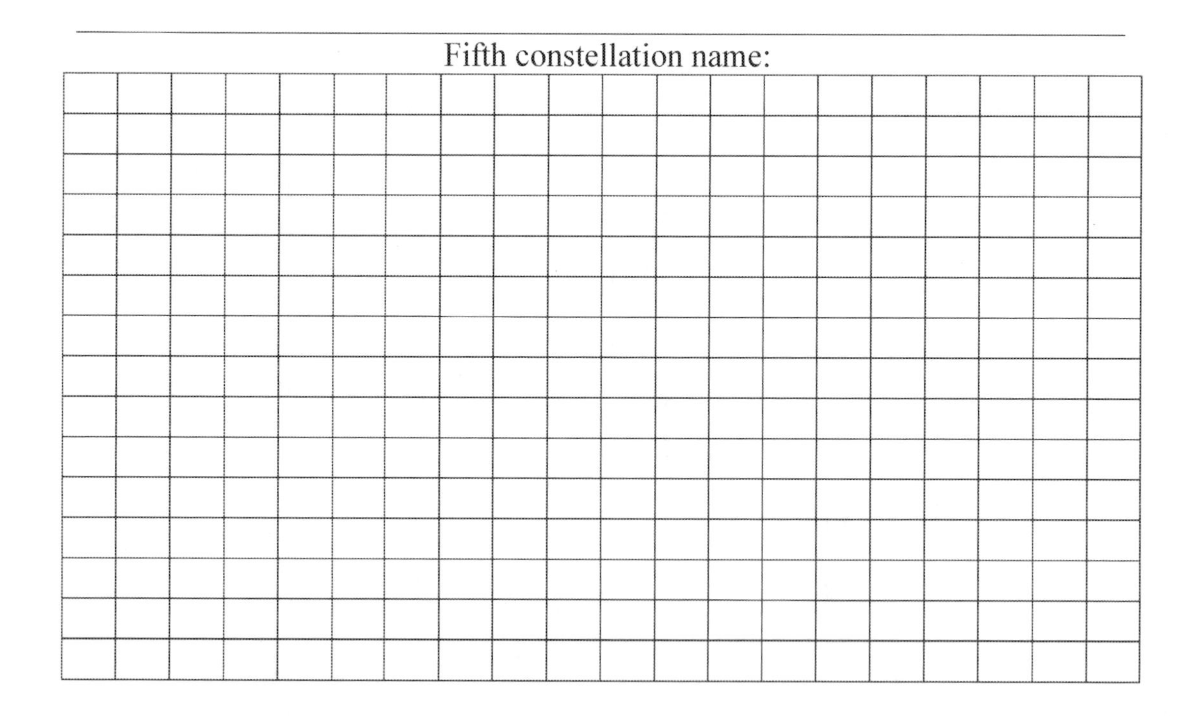

2. Making use of the star catalogue table, determine catalogue name, common name, visual magnitude and the distance from Earth for three of the brightest stars which are part of exc.1 constellations.

| Constellation 1 | Star catalog name | Star common name | Visual magnitude [mag] | Distance [pc] |
|---|---|---|---|---|
| | | | | |
| | | | | |
| | | | | |

| Constellation 2 | Star catalog name | Star common name | Visual magnitude [mag] | Distance [pc] |
|---|---|---|---|---|
| | | | | |
| | | | | |
| | | | | |

| Constellation 3 | Star catalog name | Star common name | Visual magnitude [mag] | Distance [pc] |
|---|---|---|---|---|
| | | | | |
| | | | | |
| | | | | |

| Constellation 4 | Star catalog name | Star common name | Visual magnitude [mag] | Distance [pc] |
|---|---|---|---|---|
| | | | | |
| | | | | |
| | | | | |

| Constellation 5 | Star catalog name | Star common name | Visual magnitude [mag] | Distance [pc] |
|---|---|---|---|---|
| | | | | |
| | | | | |
| | | | | |

# THE CELESTIAL SPHERE

The celestial sphere is an imaginary spherical surface centered on Earth. The projection of the terrestrial equator onto the celestial sphere gives the celestial equator. Whereas, the celestial horizon is the expansion of the terrestrial horizon onto the same sphere. Angle of inclination of the celestial equator to the celestial horizon is $90^\circ - \varphi$, where $\varphi$ is latitude.

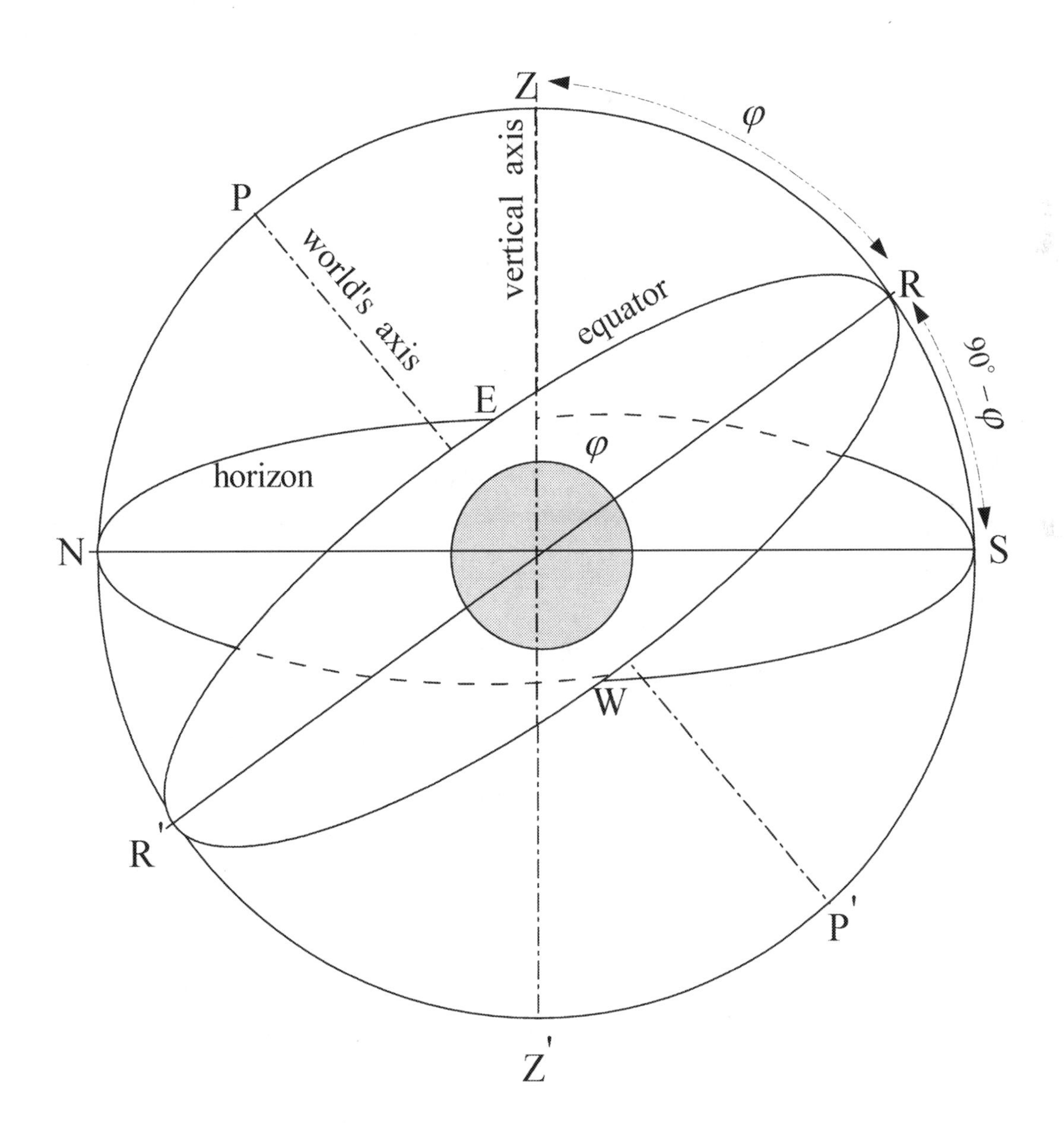


Latitude and its references on the celestial sphere

## Exercises:

1. Give three great circles perpendicular to each other on the celestial Sphere and briefly describe them.

First circle name:
Description:

Second circle name:
Description:

Third circle name:
Description:

2. Point out pairs (great circle plane – axis) on the celestial sphere which are perpendicular to each other. What principal points do they determine?

Pair 1:
Points:

________________________________________
________________________________________
________________________________________
________________________________________

Pair 2:
Points:

________________________________________
________________________________________
________________________________________
________________________________________

Pair 3:
Points:

________________________________________
________________________________________
________________________________________
________________________________________

Pair 4:
Points:

________________________________________
________________________________________
________________________________________
________________________________________

3. What is the angle between axis pointing the Northern Celestial Pole and the Point of the North? Give a proof.

# ASTRONOMICAL COORDINATE SYSTEMS

## 1) The horizon coordinate system

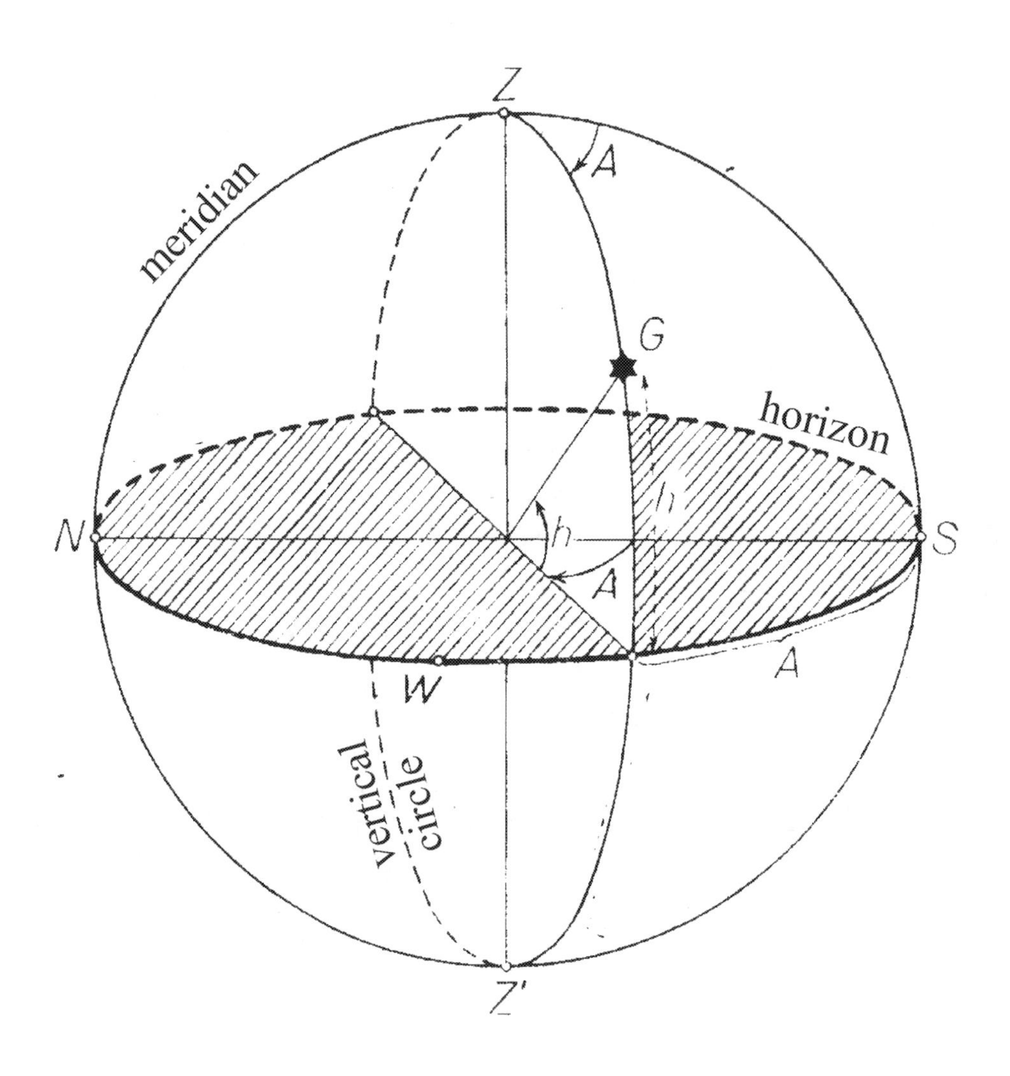


Horizon coordinate system. G – celestial body,
h – altitude, A – azimuth

**Azimuth** – the dihedral angle between the plane of the local meridian, and the plane of the vertical circle passing through a given object.

Azimuth ranges from 0° to 360° according to clockwise direction. One measures it from the point of the south.

**Altitude** – the plane angle between the astronomical horizon plane, and a given object direction. It ranges from –90° to +90°. Negative values pertain to the under-horizon objects.

### 2) The equatorial coordinate system

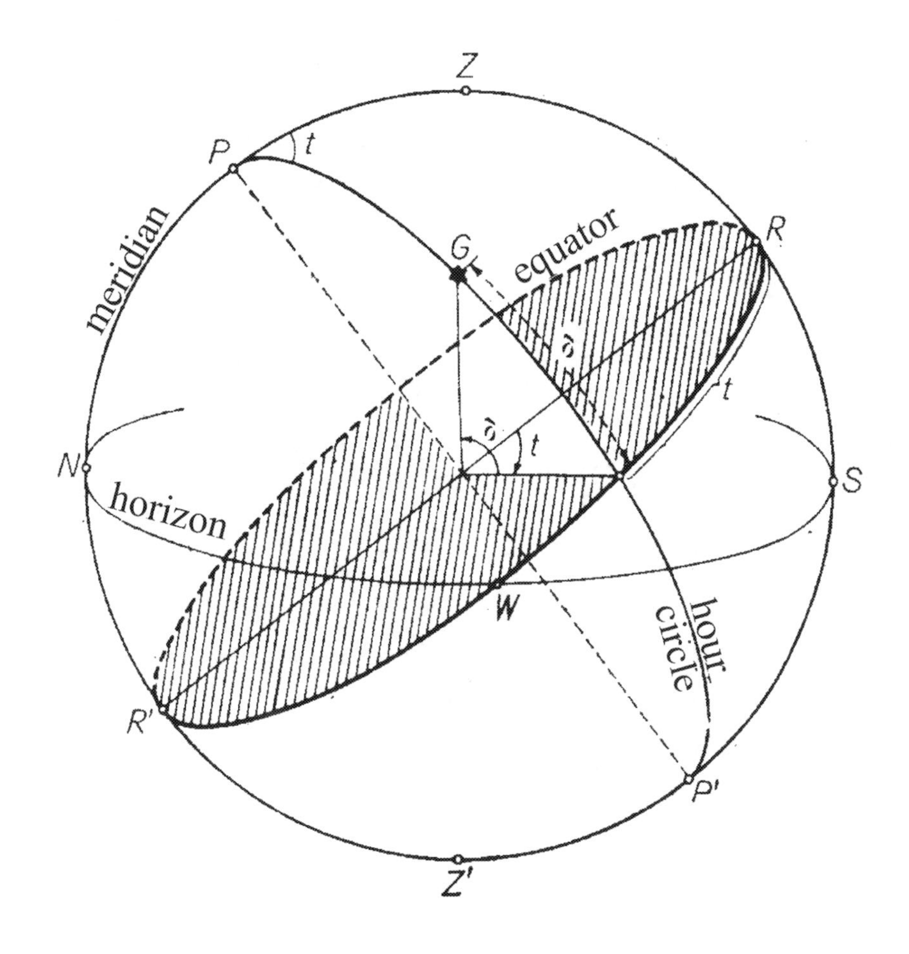


Equatorial co-ordinate system (hour system):
**δ** – declination of celestial body,
**t** – hour angle of celestial body

**Declination** – angle between the celestial equator and a given point direction. Declination ranges from 90° (northern pole) to 0° (celestial equator) to –90° (southern pole).

**Hour angle** – dihedral angle between the meridian plane and the plane of the hour circle.

The hour angle is measured accordingly to the daily motion of celestial sphere, and it ranges from 0h to 24h or 0° to 360°, where 24 hours is 360 degrees, one hour is 60 minutes and one minute is 60 arc-seconds.

### 3) The equinoctial coordinate system

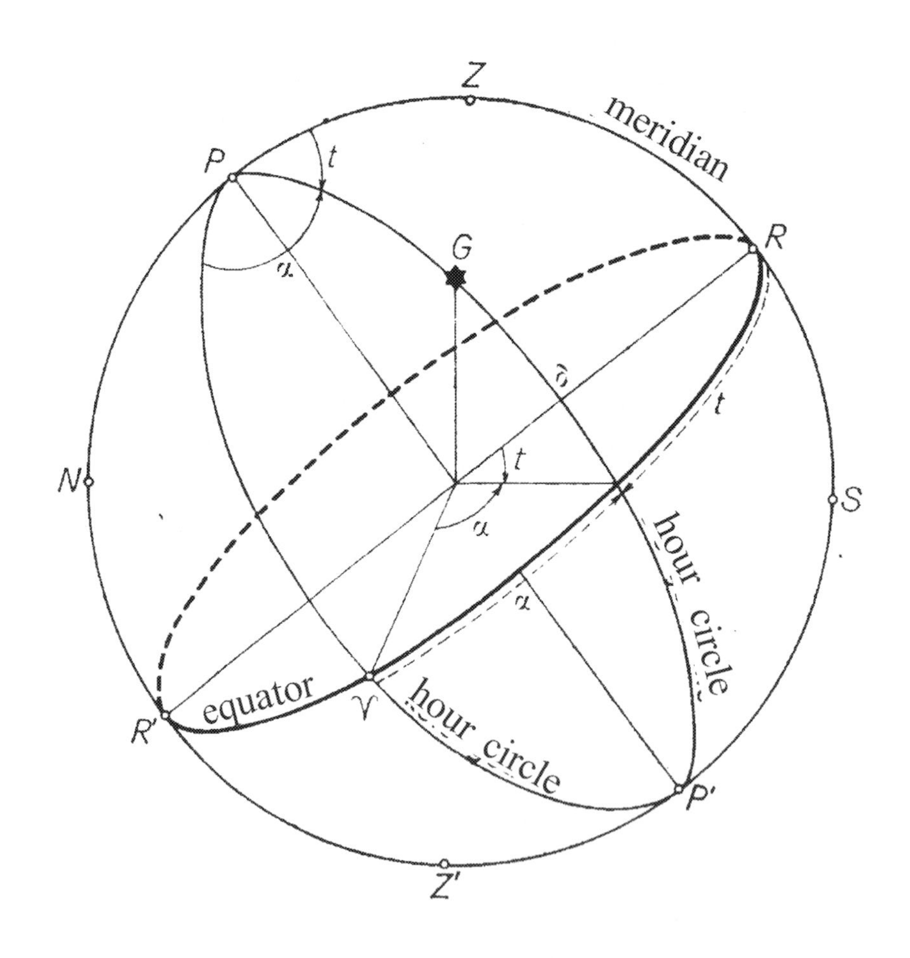


The equinoctial coordinate system in which all three coordinates have been denoted:

**α** – right ascension
**δ** – declination of celestial body,
**t** – hour angle of celestial body

**Right ascension**– dihedral angle between the plane of the hour circle of the vernal equinox and the plane of an object hour circle. The right ascension is counted eastward, according to the yearly motion of Sun. It ranges from 0h to 24h.

**Interrelation between h, φ, δ at the time of culmination**

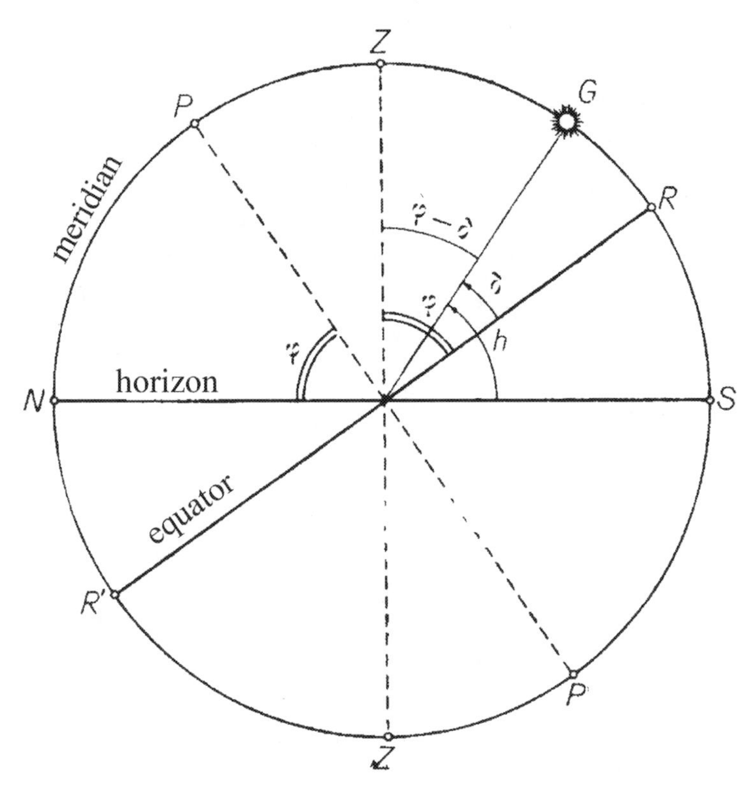


Upper culmination of a celestial body *G* (south of the zenith):

$$(\varphi-\delta)+h = 90°,$$

Upper culmination of a celestial body, when it is between zenith and the North Pole:

$$(h+\delta) - \varphi = 90^\circ$$

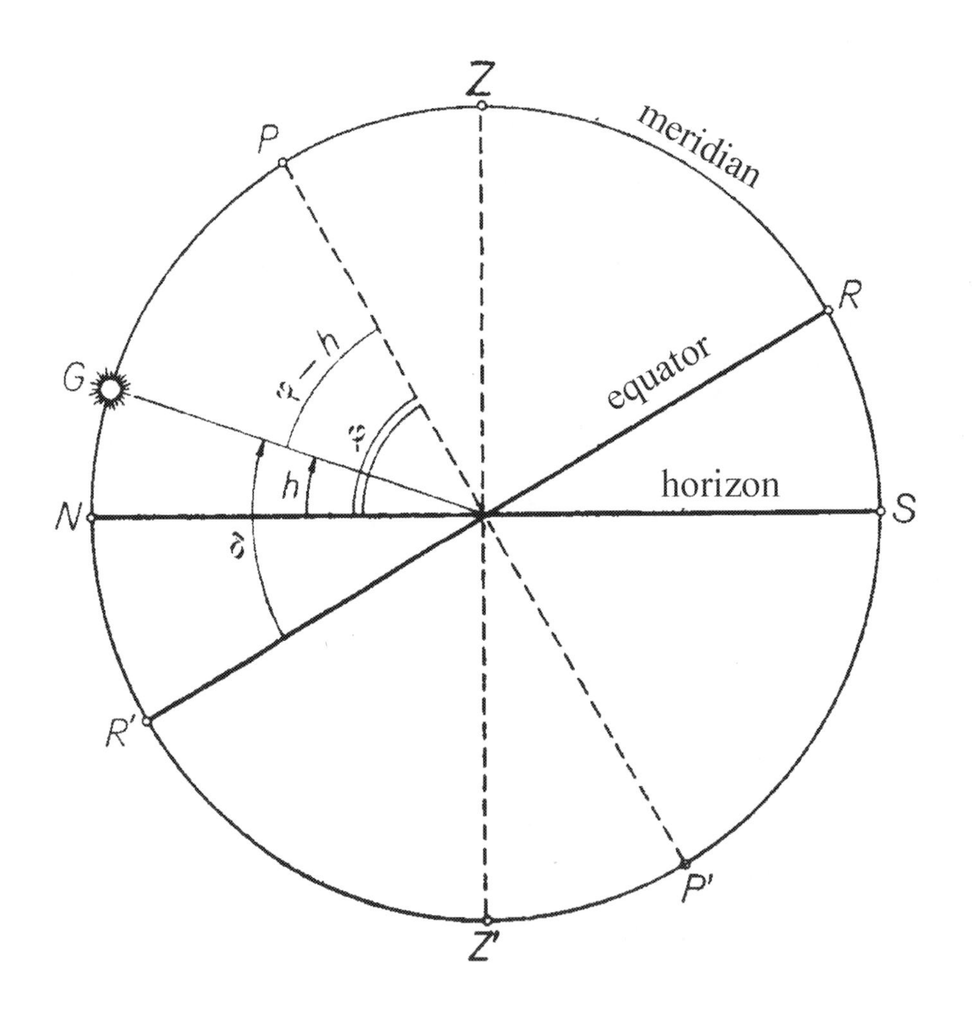


Culmination of celestial body *G* (north of zenith):

$$(\varphi - h) + \delta = 90^\circ$$

$$h = -90^\circ + \varphi + \delta.$$

Exercises:

1. Determine horizon and equatorial coordinates (using your current latitude) of:

   A. All eight known points on the celestial sphere
   B. Sun while upper culminates on March 21st

| P \ C | | | | | | | | | **Sun** |
|---|---|---|---|---|---|---|---|---|---|
| | | | | | | | | | |
| | | | | | | | | | |
| | | | | | | | | | |
| | | | | | | | | | |

2. Determine hour coordinates of the point of south and north for a place situated on the terrestrial equator.

3. Calculate altitude of the circumpolar star $\varepsilon$ *Ursae Majoris* at the upper and lower culmination. The star's declination $\delta = +56°07'$. Make calculations for $\varphi = 52°13'$.

4. Declination of the star $\alpha$ *Canis Majoris* (Sirius) is $\delta = -16°41'$. Calculate its altitude over horizon at the upper culmination ($\varphi$=51°47').

5. Calculate the Sun's altitude on June, 22nd at noon. ($\varphi$=54°31′) The Sun's declination on that day is $\delta$=+23°27′.

6. Calculate Sun's altitude at noon on June, 22nd on Tropic of Cancer. The Tropic of Cancer latitude $\varphi$ = +23°27′. Specify conclusions.

7. For a place which is located on Arctic Circle on December, 22nd, $\varphi$ = +63°33′ and declination $\delta$ = –23°27′. Determine the Sun's altitude at noon. Specify conclusions.

# THE MOON

The moon is the biggest and the brightest object in the night sky, though the moon itself is not a light source; rather, it reflects the Sun's rays. The Earth – Moon distance is 384,000 km. As the Moon revolves around Earth, differing amounts of sunlight are reflected, causing the familiar phases of the Moon. The New phase is a day one, whereas the Full moon can be observed the whole night. You can trace the waxing moon in the evening the best, and the waning one – after midnight.
The time needed for the moon to make one complete revolution around Earth, and take the same position among the stars as its initial one, is called the sidereal period. In order to obtain the same phase, the moon needs more time. Therefore, we also determine the synodic period.
To find angular size of an object or its linear diameter you can use the small-angle formula.

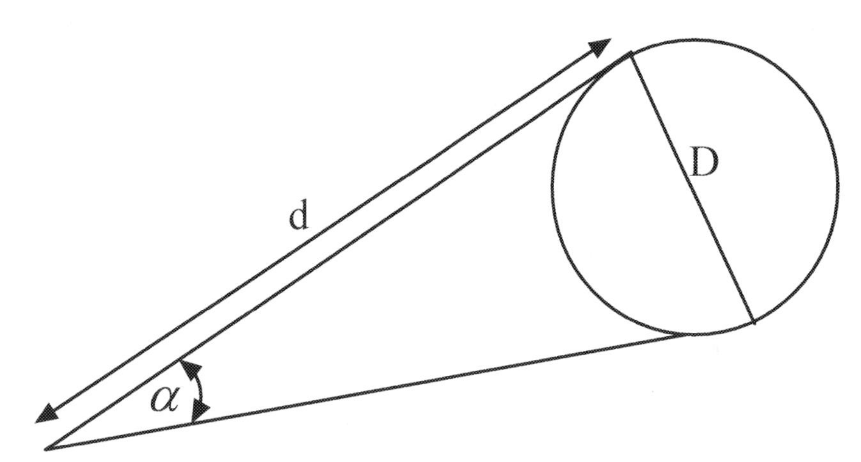


$$\frac{d}{D} = \frac{\varphi}{\alpha}$$

$D$ – *linear diameter*

$d$ – *distance to an object*

$\alpha$ – *angular size*

$\varphi$ – *angle corresponding to 1 rad*

Exercises:

1. The angular diameter of the moon is approximately $0.5^\circ$, and its distance to Earth is 384,000 km. Apply the small-angle formula to calculate its linear diameter.

2. Given the following phases of the moon, what would its position be, expressed in degrees?

| Phase | Position (in degrees) |
|---|---|
| New | |
| Waxing gibbous | |
| Last quarter | |
| Waxing crescent | |
| Waning gibbous | |
| Full | |
| Waning crescent | |
| First quarter | |

3. How many days on average elapse between each neighboring phase pair given in problem #2?

| Pair | Time (days) |
|---|---|
| New – Waxing gibbous | |
| | |
| | |
| | |
| | |
| | |
| | |

# THE SUN

## Dimension of the sun

To assess the linear dimensions of the Sun, it is enough to measure its angular diameter. The diameter changes in the course of the Earth's yearly motion, but it assumes the average angular diameter $\alpha = 31'59''$. According to the drawing, one can write:

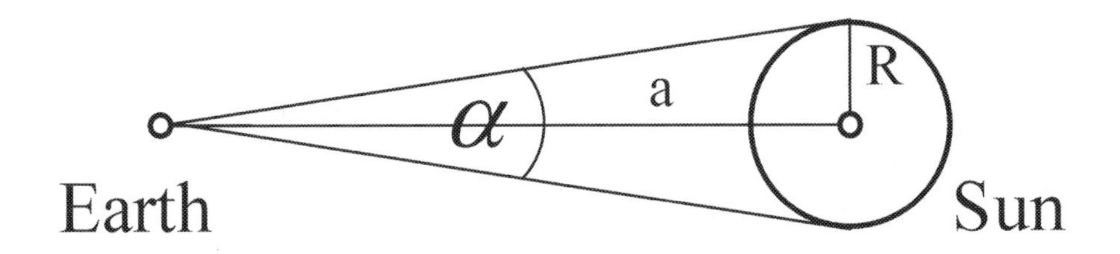


$$R = a\sin\left(\frac{\alpha}{2}\right)$$

$$d = 2R$$

The spectral analysis of the Sun's light allows us to determine its chemical composition. The solar spectrum consists of the continuous emission spectrum and the dark lines of the so called absorption spectrum. This system of characteristic lines enables us to indicate what chemical elements comprise the Sun, and more than 70 elements have been discovered using this method.

There is a list of some elements in the table below along with their absorption lines which occur in the sun's spectrum.

| Wavelength [nm] | Fraunhofer denotation | Element | Color |
|---|---|---|---|
| 656.282 | C | H ($H_\alpha$) | red |
| 589.594 | $D_1$ | Na | yellow |
| 588.998 | $D_2$ | Na | yellow |
| 518.362 | $b_1$ | Mg | green |
| 517.270 | $b_2$ | Mg | green |
| 516.733 | $b_4$ | Mg | green |
| 486.134 | F | H ($H_\beta$) | blue |
| 438.350 | d | Fe | blue |
| 434.048 | G' | H ($H_\gamma$) | violet |
| 422.674 | g | Ca | violet |
| 410.175 | h | H ($H_\delta$) | violet |
| 406.361 | - | Fe | violet |
| 404.583 | - | Fe | violet |
| 396.849 | H | $Ca^+$ | violet |
| 393.368 | K | $Ca^+$ | violet |
| 382.044 | L | Fe | violet |

Exercises:

1. Given the mean Earth – Sun distance of 149,600,000 km, calculate how many times the Sun's diameter is bigger than the Earth's. Take the Earth's radius to be 6370 km.

Data: Unknowns:

Solution:

2. In the sun's colorful spectrum the Fraunhofer lines have been detected. Determine what scope of colors are contained between two picked lines. What is the average wavelength difference between them?

| **Lines** | **Colors** | **Wavelength difference [nm]** |
|---|---|---|
| $D_2 \div G'$ | yellow, green, blue, violet | 154.95 |
| | | |
| | | |
| | | |

3. Draw a colorful Sun's spectrum, and place lines pertaining to at least three chemical elements. Name these elements on the spectrum.

# H – R DIAGRAM

The H – R diagram classifies stars: it gives the relative luminosity of a star versus temperature of the star's photosphere. The diagram can be applied also with brightness and spectral type of the star. Stars on the diagram are grouped into certain areas. The majority of stars make up the main sequence: extending along the diagonal from the upper-left corner to the lower-right one in the diagram. The branch which lies directly under the main sequence is the subdwarfs branch. A group of white dwarfs is at the bottom-left diagram, whereas subgiants, red giants, giants and supergiants are above the sequence.

Spectral lines of observed stars depend on their atmosphere density. The lower its density, the sharper the lines are.

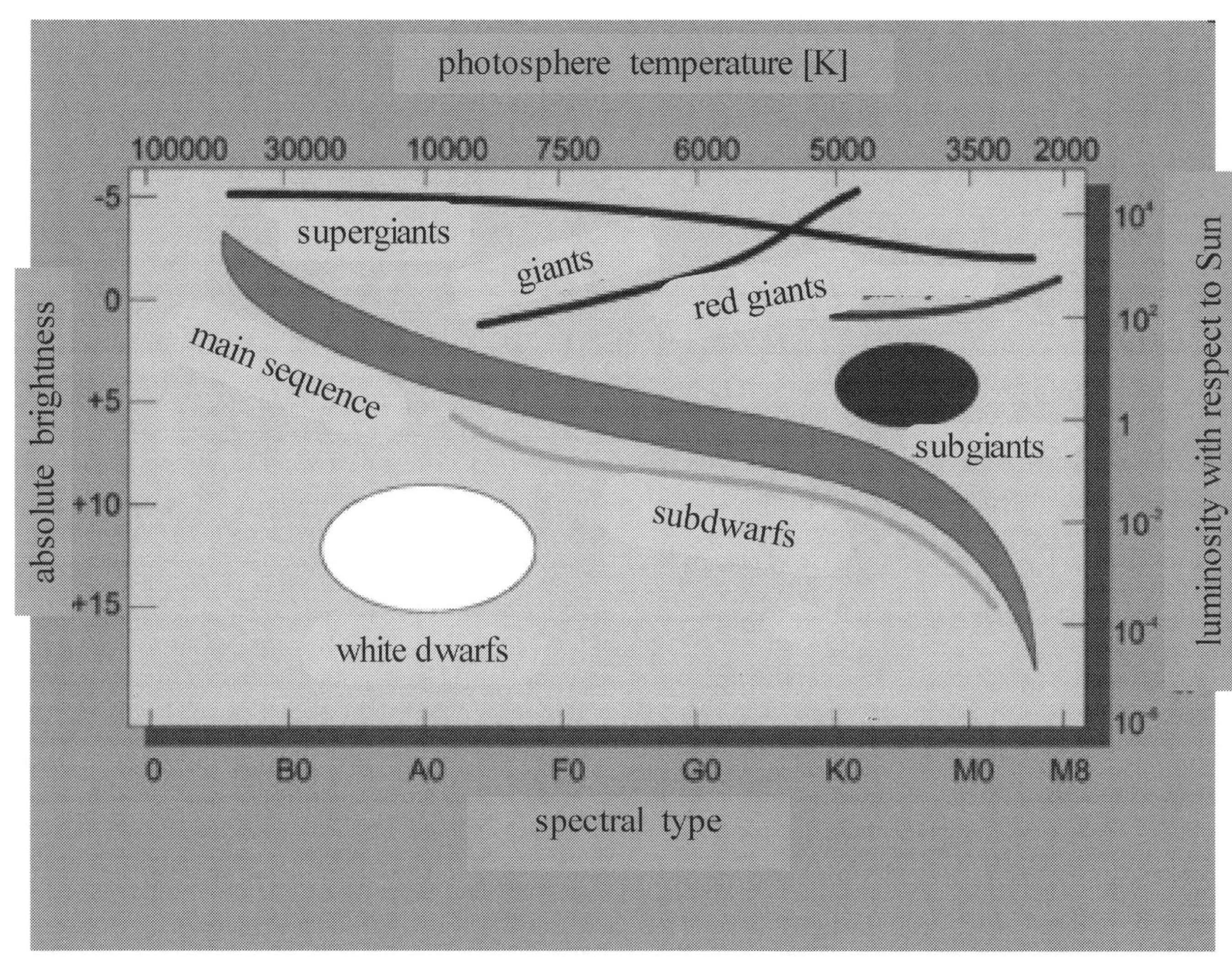

Luminosity classes:

- V: Main sequence stars
- IV: Subgiants
- III: Giants
- II: Bright giants
- Ib: Supergiants
- Ia: Bright supergiants
- Ia-0: Hypergiant
- D: White dwarfs

## Exercises:

1. Below the given table, draw a coordinate system corresponding to the H-R diagram, and plot the location of these stars.

| Star | Type | Absolute Magnitude |
|---|---|---|
| α Cas | K0 | –1.985 |
| β And | M0 | –1.76 |
| δ Per | B5 | –3.04 |
| ε Per *A* | B0.5 | –3.19 |
| δ Cas | A5 | +2.1 |
| β Ori *A* | B8 | –7.84 |
| δ CMa | F8 | –6.87 |
| γ Ori | B2 | –2.72 |
| η Gem *A* | M3 | –2 |
| τ Pup | K0 | +0.1 |
| o UMa *A* | G5 | –0.4 |
| κ Vel | B2 | –3.4 |
| θ UMa *A* | F6 | +2.43 |
| ε Cen | B1 | –3.9 |
| π Her | K3 | –2.1 |
| β Dra *A* | G2 | –2.28 |
| α Cyg | A2 | –8.38 |
| β Aqr | G0 | –3.34 |
| ε Peg *A* | K2 | –4.14 |

Diagram:

2. Making use of the *star-type-absolute magnitude* table above, complete the following table:

| **Star** | **Average temperature of photosphere (T) [K]** | **Average relative luminosity($L/L_0$)** | **Luminosity class** |
|---|---|---|---|
| α Cas | 4800 | $10^2$ | II – III |
| | | | |
| | | | |
| | | | |
| | | | |
| | | | |
| | | | |
| | | | |
| | | | |
| | | | |
| | | | |
| | | | |
| | | | |
| | | | |
| | | | |
| | | | |
| | | | |
| | | | |
| | | | |

3. Group the stars from the *star-type-absolute magnitude* table according to luminosity classification. Draw a schematic spectrum showing how the width of spectral lines depends on the luminosity class.

## BASIC QUANTITIES CHARACTERIZING STARS

**Apparent brightness** (apparent magnitude(m)) – the amount of a star's energy reaching Earth. It is dependent on both the real brightness and the distance to Earth.

**Absolute brightness** (absolute magnitude (M)) – the measure of energy emitted by a star out into the surrounding space. It corresponds to a real brightness at a distance of 10 pc.

Apparent brightness ~ Flux (f) $\left[\frac{J}{s \cdot m}\right]$

$$m_1 - m_2 = -2.5\log\left(\frac{f_1}{f_2}\right)$$

$$M = m + 5 - 5\log d$$

$$d = 10^{(m+5-M)/5}$$

**Luminosity** – the total energy a star gives off in a unit of time

$$L = 4\pi R^2 \sigma T^4$$

***R*** - radius
***T*** - temperature
σ – Stefan-Boltzmann constant
***L*** - luminosity

$$\frac{L}{L_S} = \left(\frac{R}{R_S}\right)^2 \left(\frac{T}{T_S}\right)^4$$

# Mass – Luminosity Relation (main sequence stars)

$$L = M^{3.5}$$

Exercises:

1. Complete the following table:

| Star | Visual apparent magnitude | Distance [ly] |
|---|---|---|
| α Cas | 2.22 | |
| δ Per | | 590 |
| β Ori *A* | 0.15 | |
| o UMa *A* | | 150 |
| α Cyg | 1.25 | |

2. Find the relative luminosity $L/L_s$ ($L_s$ – Sun's luminosity) of a star given its relative radius $R/R_s$ ($R_s$ – Sun's radius) and relative temperature $T/T_s$ ($T_s$ – temperature of the Sun's photosphere).

I.

$R/R_s = 100$, $T/T_s = 1/3$

Calculations:

II.

$R/R_s = 1$, $T/T_s = 10^3$

Calculations:

III.

$R/R_s = 10^4$, $T/T_s = 1/10^2$

Calculations:

3. Plot a graph of luminosity (in terms of the Sun's luminosity) as a function of mass (in solar masses) for main sequence stars. Begin with stars whose mass is a fraction of the Sun's, and finish with stars whose mass is several hundred times greater than the mass of the sun.

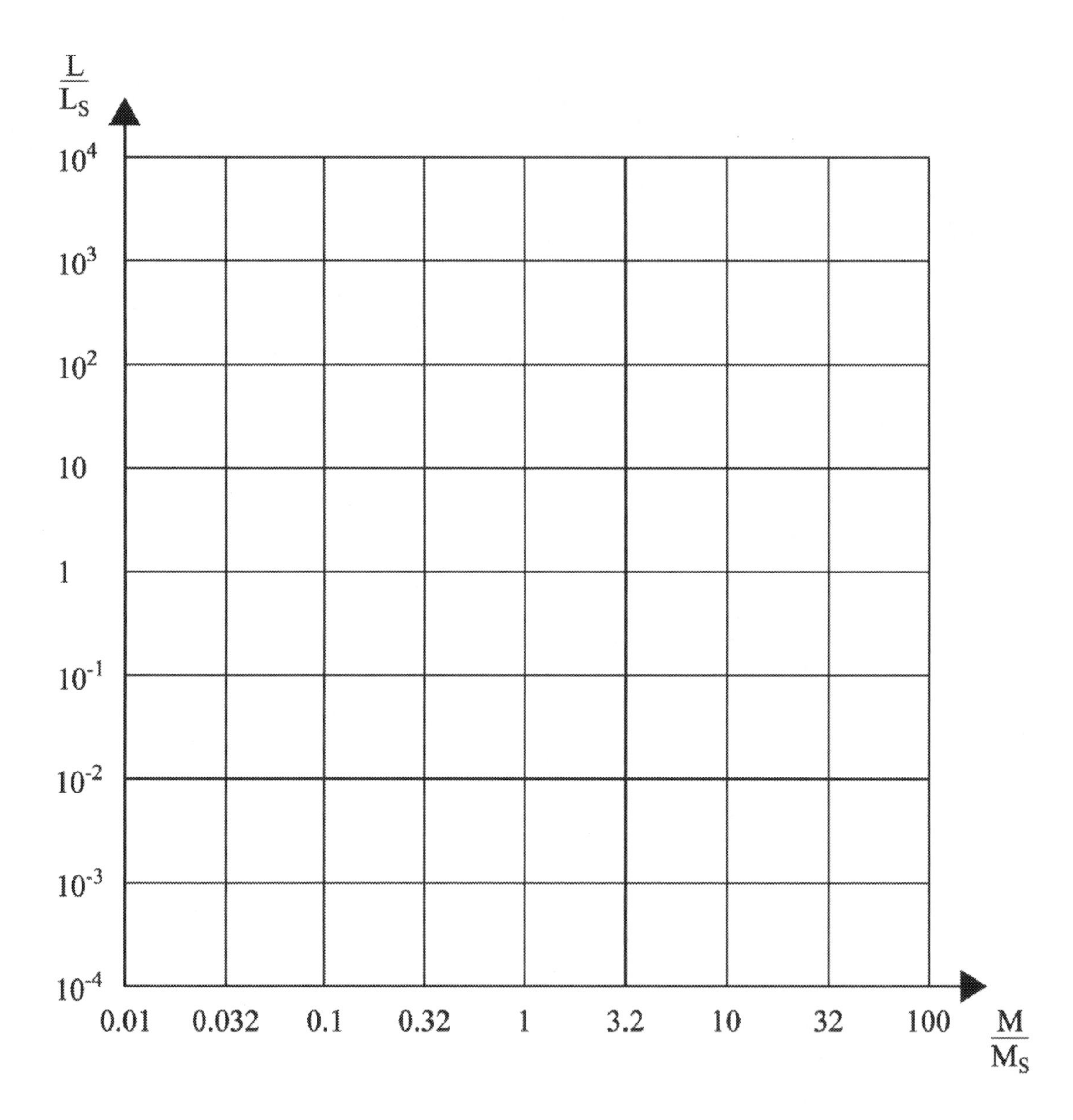

# VARIABLE STARS

Variable stars change their brightness within certain periods of time or irregularly. One can plot graphs of variability as brightness as a function of time. Such a graph, called a light curve, allows us to determine a series of quantities depicting stars. In general, one may distinguish two main kinds of variable stars: eclipsing binaries and intrinsic variables that change their brightness due to internal processes.

Brightness curvature of the X Cam star is shown below. The period is about 145 days.

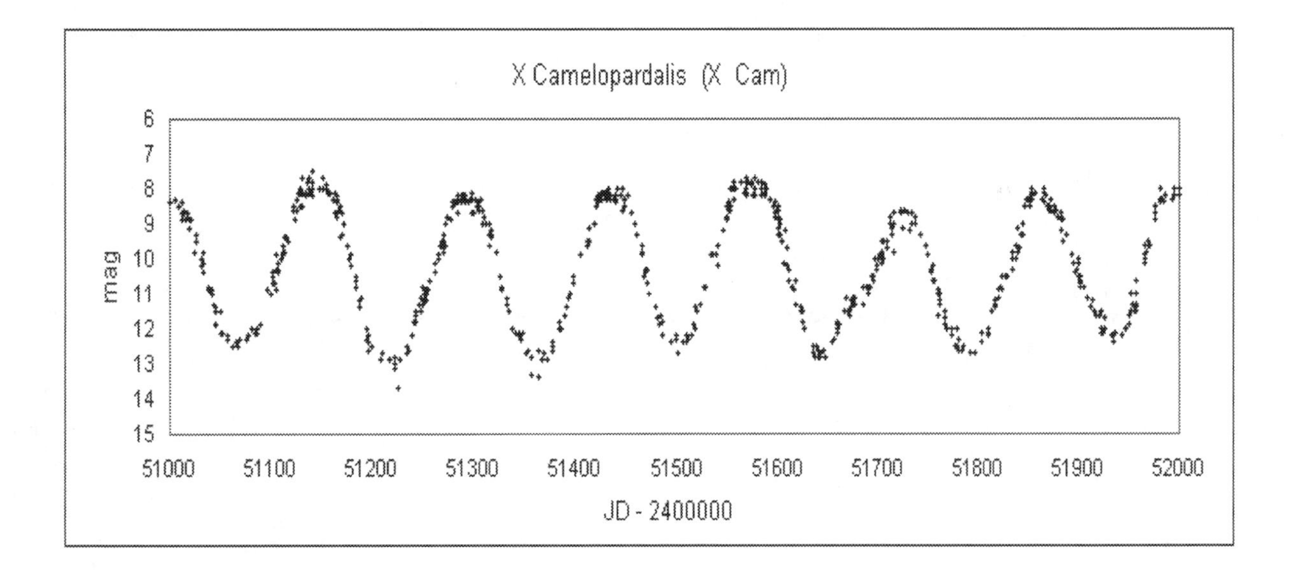

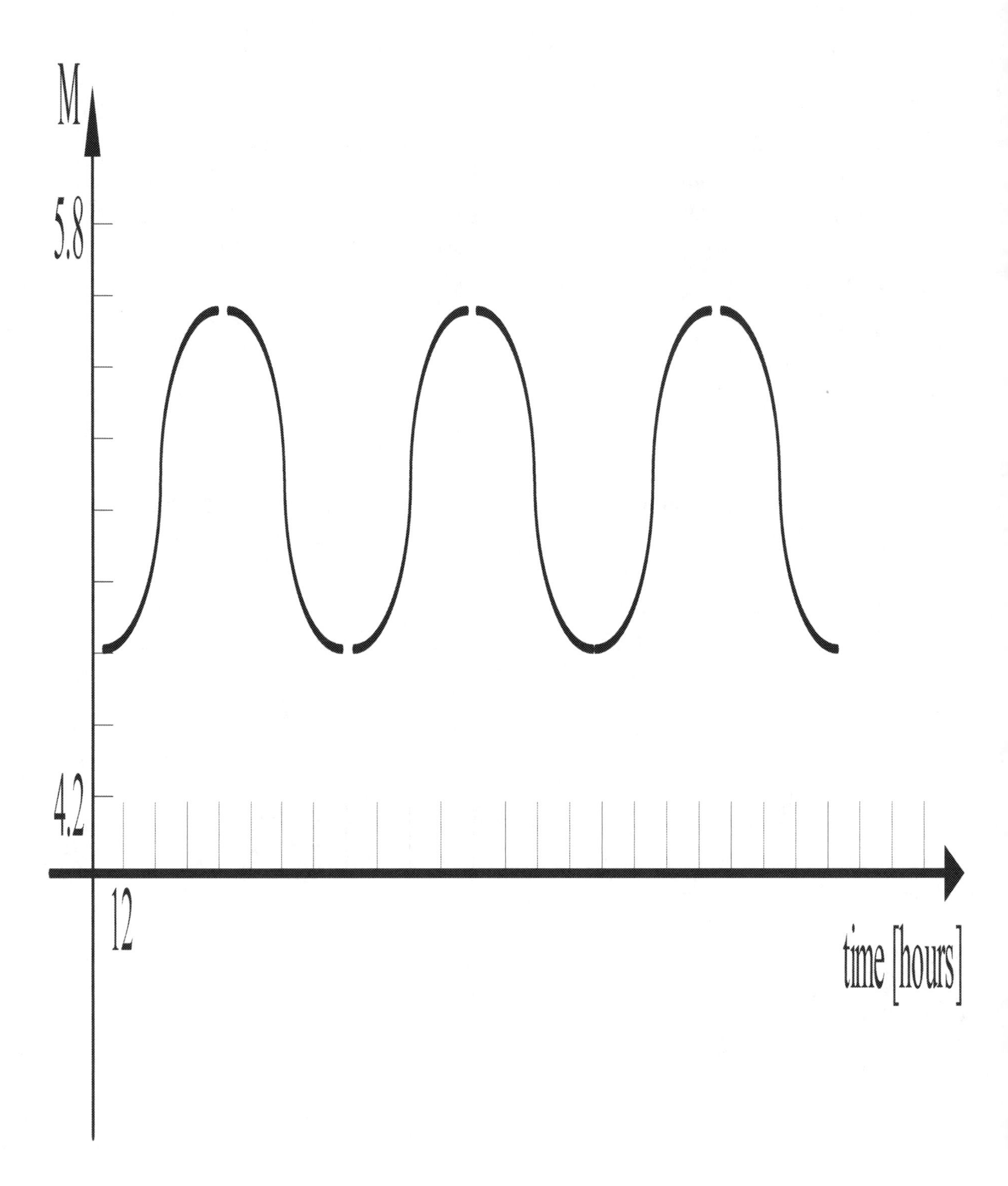


Figure – magnitude versus time

Exercises:

1. Using the above *Figure-magnitude versus time* of the given star, determine its period of variability, frequency and magnitude range.

Period:
----------------------------------------------------------------------------------------------------

Frequency:
----------------------------------------------------------------------------------------------------

Magnitude range:
----------------------------------------------------------------------------------------------------

2. What is its spectral type? ____________________________________________

3. What is the range of its surface temperature?________________________

4. What should its average luminosity be (with respect to the sun)?

____________________________________________________________________

# THE HUBBLE CONSTANT

The Hubble law is a basic law of observational cosmology that links the distance to galaxies ***d*** with their recession velocities ***v*** :

$$v = H \cdot d$$

where ***H*** is the Hubble constant. The recession velocity measure is the redshift. This relation is true for galaxies which are relatively close to us. Hubble's law states that the greater the distance to a galaxy the greater its recession velocity and therefore yielding a larger redshift. Analogously, Doppler's law can be used to calculate redshift when any galaxy is receding with respect to others; specifically, when a light source is receding with respect to the observer, then Doppler's law takes the form:

$$f_0 \approx f_s \frac{c - v}{c} \approx f_s \frac{c}{c + v}$$

$f_0$ – the wave frequency that is received by observer

$f_s$ – the wave frequency that is generated by the source

$c$ – the speed of light in a vacuum

$v$ – the wave speed

Keeping in mind the wavelength – frequency relation $\lambda = \frac{v}{f}$, one can give the wavelength dependent form of Doppler's law.

Exercises:

1. The spectrometric investigation of a distant galaxy's radiation has been carried out. The spectroscopic line of wavelength $\lambda$ = 730 nm has been identified as the Balmer series hydrogen line, with an observed wavelength at the Earth laboratory of $\lambda_0$ = 487 nm. Making use of Doppler's formulas, determine the speed and the direction of the galaxy's motion.

2. The observations indicate that the universe expands homogeneously and does not have a distinct center. For galaxies to which the Hubble law is applicable, determine:
   a. In what way the ratio $\lambda/\lambda_0$ depends on the source – Earth distance? Derive the Hubble constant as a function of $d$ and $\lambda/\lambda_0$.

I. $\lambda/\lambda_0 - d$ interrelation:

II. The Hubble constant derivation:

b. In what way does redshift (defined as: $z = \frac{\lambda - \lambda_0}{\lambda_0} = \frac{\Delta\lambda}{\lambda_0}$) depend on the source – Earth distance?

c. Calculate the Hubble constant for

(i) the galaxy in the constellation Virgo $d_1 = 17$ Mpc, $z_1 = 0.004$

(ii) the galaxy in the constellation Ursa Major $d_2 = 180$ Mpc, $z_2 = 0.051$

(i)

(ii)

# THE AGE OF THE UNIVERSE

We can calculate the age of the universe given the mean of times for many galaxies:

$$T_i = \frac{d_i}{v_i}$$

where $d_i$ - the distance between our Galaxy and a galaxy of number "$i$ ", $v_i$ - the velocity of "$i$-th " galaxy. The age of the universe can be estimated as:

$$T = \frac{\sum_{i=1}^{n} T_i}{n}$$

where "$n$" is the number of galaxies that have been taken into account.

| Galaxy | Distance |
|---|---|
| Milky Way | 0 |
| Omega Centauri | 18 kly |
| Canis Major Dwarf | 25 kly |
| Virgo Stellar Stream | 30 kly |
| Sagittarius Dwarf | 80 kly |
| Large Magellanic Cloud | 160 kly |
| Small Magellanic Cloud | 200 kly |
| Andromeda | 2.5 Mly |
| Triangulum | 2.9 Mly |

Exercises:

1. Making use of the above table and the Hubble law, calculate the velocity of each tabled galaxy, and next, the time of its life.

| Galaxy | Velocity | Life's Time |
|---|---|---|
| Milky Way | | |
| Omega Centauri | | |
| Canis Major Dwarf | | |
| Virgo Stellar Stream | | |
| Sagittarius Dwarf | | |
| Large Magellanic Cloud | | |
| Small Magellanic Cloud | | |
| Andromeda | | |
| Triangulum | | |

2. Plot a graph of distance versus velocity for the above galaxies.

d

v

3. Determine the age of the universe. Subsequently, compare the found result with that given by the formula: $T = \frac{1}{H}$, where $H$ is the Hubble constant. Assess the standard deviation.

Calculations:

Comparison & Standard Deviation:

REMARKS & NOTES